T0350215

Introduction to Geometric Probability

This is the first modern introduction to geometric probability, also known as integral geometry. The subject is presented at an elementary level, requiring little more than first year graduate mathematics. The theory of intrinsic volumes due to Hadwiger, McMullen, Santaló and others is presented, along with a complete and elementary proof of Hadwiger's characterization theorem of invariant measures in Euclidean n-space. The theory of the Euler characteristic is developed from an integral-geometric point of view. The authors then prove the fundamental theorem of integral geometry, namely the kinematic formula. Finally the analogies between invariant measures on polyconvex sets and measures on order ideals of finite partially ordered sets are investigated. The relationship between convex geometry and enumerative combinatorics motivates much of the presentation. Every chapter concludes with a list of unsolved problems. Geometers and combinatorialists will find this a stimulating and fruitful tale.

Daniel A. Klain is Assistant Professor of Mathematics at Georgia Institute of Technology.

Gian-Carlo Rota is Professor of Applied Mathematics and Philosophy, Massachusetts Institute of Technology.

Lezioni Lincee
Sponsored by *Foundazione IBM Italia*
Editor: Luigi A. Radicati di Brozolo, Scuola Normale Superiore, Pisa

This series of books arises from lectures given under the auspices of the Accademia Nazionale dei Lincei and is sponsored by *Foundazione IBM Italia*.
The lectures, given by international authorities, will range on scientific topics from mathematics and physics through to biology and economics. The books are intended for a broad audience of graduate students and faculty members, and are meant to provide a '*mise au point*' for the subject with which they deal. The symbol of the Accademia, the lynx, is noted for its sharp-sightedness; the volumes in this series will be penetrating studies of scientific topics of contemporary interest.

Already published

Introduction to Geometric Probability

DANIEL A. KLAIN

GIAN-CARLO ROTA

CAMBRIDGE
UNIVERSITY PRESS

CAMBRIDGE UNIVERSITY PRESS
Cambridge, New York, Melbourne, Madrid, Cape Town, Singapore,
São Paulo, Delhi, Dubai, Tokyo

Cambridge University Press
The Edinburgh Building, Cambridge CB2 8RU, UK

Published in the United States of America by Cambridge University Press, New York

www.cambridge.org
Information on this title: www.cambridge.org/9780521593625

First published 1997
Reprinted 1999

A catalogue record for this publication is available from the British Library

ISBN 978-0-521-59362-5 Hardback
ISBN 978-0-521-59654-1 Paperback

Transferred to digital printing 2010

Contents

Preface

If we were allowed to rename the field of geometric probability – sometimes already renamed integral geometry – then we would be tempted to choose the oxymoron 'continuous combinatorics.' On more than one occasion the two fields, geometric probability and enumerative combinatorics, are brought together by mathematical analogy, that most effective breaker of barriers.

Like combinatorial enumeration, where sequences of objects bearing a common feature are unified by the idea of a generating function, geometric probability studies sets of geometric objects bearing a common feature, which are unified by the idea of an invariant measure. The basic idea is extremely simple. When considering straight lines, pairs of points, or triangles in space, one determines the invariant measure on the variety of straight lines, of pairs of points, of triangles. This idea is strangely reminiscent of the underlying idea of enumerative geometry, with one major difference: whereas enumerative geometry is bound to the counting of finite sets, geometric probability is given greater freedom, by extending the concept of enumeration to allow the assigning of invariant measures. Invariant measures are far easier to compute and, we dare add, more useful than the curiously large integers that are computed in enumerative geometry. This basic idea goes back to Crofton's article in the ninth edition of the Encyclopaedia Britannica, an article that created the subject from scratch and that is still worth reading today. The one other brilliant contribution to geometric probability in the past century was Barbier's solution of the Buffon needle problem, which remains to this day the basic trick of the subject, still being secretly exploited in ever unsuspected ways.

Geometric probability has suffered in this century the fate of other fields that would have enjoyed a healthy autonomous development, had it not been for the overpowering development of representation theory. One can reduce integral geometry to the study of actions of Lie groups, to symmetric spaces, to the Radon transform; in so doing, however, the authentic problematic of the subject is lost. Geometric probability is a customer of representation theory, in the same sense that mechanics is a customer of the calculus.

The purpose of this book is to present the three basic ideas of geometric probability, stripped of all reliance on group-theoretic techniques. First, we investigate measures on polyconvex sets (i.e., finite unions of compact convex sets) in Euclidean spaces of arbitrary dimension that are invariant under the group of Euclidean motions. A great many mathematicians are still basking in the illusion that there is only one such measure, namely, the volume. We merrily destroy this illusion by proving what is at present the fundamental result of the field (due to Hadwiger), stating that the space of such invariant measures is of dimension $n + 1$ in a Euclidean space of dimension n. The proof of this fundamental result given in the text is new, due to the first author. It becomes clear, on reading the applications of the fundamental theorem, that the basic invariant measure to be singled out from such a bounty is not the volume, but the Euler characteristic (as Steve Schanuel was first to realize). Here again we meet with wide ignorance on the part of the mathematical public: the fundamental fact that the Euler characteristic is an invariant measure (in fact, it is the only integer-valued invariant measure) is not as well known as it should be. It leads to one-line proofs of most of the fundamental theorems on convex sets. We develop the theory of the Euler characteristic from scratch, in a way that makes it look like an ordinary integral.

Second, we prove the fundamental formula of integral geometry, viz., the kinematic formula. Here we displace the common device of Minkowski sums from its typically central role, not merely as a display of mathematical machismo, but with an ulterior motive.

Third, we try to bring out from the beginning the striking analogy between the computation of invariant measures and certain combinatorial properties of finite partially ordered sets. The second author pointed out in 1967 that the notion of Euler characteristic could be extended to such partially ordered sets by means of the Möbius function. We now go one step further and show that an analogue of the theory of invariant measures in Euclidean space can be worked out in partially ordered sets, including finite analogues of the kinematic formula and even of Helly's theorem. This analogy brings out in stark contrast the unexplored terrain of classical geometric probability, namely, a thorough understanding of the integral geometric structure of the lattice of subspaces of Euclidean space under the action of the orthogonal group. It also brings us closer to the current outer limits of mathematics, to the theory of Hecke algebras, to Schubert varieties and to the quantum world.

We hope that the reading of this introduction to the field of geometric probability will encourage further development of these analogies.

The text is based on the 'Lezioni Lincee' given by the second author in 1986 at the Scuola Normale Superiore in Pisa. The authors wish to thank Ennio De Giorgi, Edoardo Vesentini, and Luigi Radicati for providing an interested audience for the original lectures. Thanks are also due to Stefano Mortola for his careful reading of the initial draft, and to Beifang Chen, Steve Fisk, Joseph Fu, Steven Holt, Erwin Lutwak, and three anonymous referees for their valuable comments and suggestions.

Using this book

Although parts of this book assume a knowledge of basic point-set topology, measure theory, and elementary probability theory, the greater part of the text should be accessible to advanced undergraduates. Proofs are either given in full or else stretched to the point from which the reader will be able to reconstruct them without effort. Only on certain technical measure-theoretic points have we felt the need to omit details that, although indispensable in a detailed treatment, are of questionable relevance in an exposition that is meant to stress geometric insight and combinatorial analogy. Some notions that appear vague in the early sections will be revisited later on, after language has been developed for a treatment in clear and rigorous terms. References and open problems are deferred to the notes at the end of each chapter.

1

The Buffon needle problem

We begin with what is probably the best-known problem of geometric probability, the Buffon needle problem. This solution of the needle problem via the characterization of an additive set functional serves to motivate the study of valuations on lattices, the topic of Chapter 2. Variations and generalizations of the Buffon needle problem are presented in Chapters 8 and 9.

1.1 The classical problem

Parallel straight lines are drawn on the plane $\mathbf{R}^2$, at a distance d from each other. A needle of length L is dropped at random on the plane. What is the probability that the needle shall meet at least one of the lines?

This problem can be solved by computations with conditional probability (Feller, for example, solved it in this way in his well known treatise [23, p. 61]). It is, however, more instructive to solve it by another method, one that minimizes the amount of computation and maximizes the role of probabilistic reasoning.

Let X_1 be the number of intersections of a randomly dropped needle of length L_1 with any of the parallel straight lines. If the needle is long enough, the random variable X_1 can take several integer values, whereas if the needle is short, it can take only the values 0 or 1.

If p_n is the probability that the needle meets exactly n of the straight lines, and if $E(X_1)$ denotes the expectation of the random variable X_1, then we have

$$E(X_1) = \sum_{n \geq 0} n p_n.$$

Thus, if $L_1 < d$, then

$$E(X_1) = 0p_0 + 1p_1 = p_1,$$

and p_1 is the probability we seek. Therefore, it is sufficient to compute the expectation $E(X_1)$. Suppose that another needle of length L_2 is dropped at random. The number of intersections of this second needle with any of the parallel straight lines drawn on $\mathbf{R}^2$ is another random variable, say X_2. The random variables X_1 and X_2 are independent, unless the needles are welded together. Suppose that the needles are rigidly bound at one of their endpoints. They may form a straight line, or they may be at an angle. In either case, if the two rigidly bound needles are simultaneously dropped on $\mathbf{R}^2$, their total number of intersections will still be $X_1 + X_2$. The random variables X_1 and X_2 will no longer be independent, but their expectation will remain additive:

$$E(X_1 + X_2) = E(X_1) + E(X_2). \tag{1.1}$$

The same reasoning applies to the random variable $X_1 + X_2 + \cdots + X_k$, for the case in which k needles are welded together to form a polygonal line of arbitrary shape.

Since $E(X_1)$ clearly depends on the length L_1, we can write $E(X_1) = f(L_1)$, where f is a function to be determined. By welding together two needles so that they form one straight line we find that $E(X_1 + X_2) = f(L_1 + L_2)$, and we infer from (1.1)

$$f(L_1 + L_2) = f(L_1) + f(L_2).$$

It then follows that f is linear when restricted to rational values of L. Since f is clearly a monotonically increasing function with respect to L, we infer that $f(L) = rL$ for all $L \in \mathbf{R}$, where the constant r is to be determined.

If C is a rigid wire of length L, dropped randomly on $\mathbf{R}^2$, and if Y is the number of intersections of C with any of the straight lines, then C can be approximated by polygonal wires, so that Y is approximately equal to $X_1 + X_2 + \cdots + X_k$. Passing to the limit, we find that

$$E(Y) = rL. \tag{1.2}$$

This allows us to determine the value of the constant r, by choosing a wire of suitable shape. Let C be a circular wire of diameter d. Obviously $E(Y) = 2$, and $L = \pi d$. It then follows from (1.2) that

$$2 = r\pi d,$$

whence $r = 2/(\pi d)$. Thus, for a short needle, we have

$$E(X_1) = p_1 = \frac{2L}{\pi d}.$$

This result has been used (rather inefficiently) to compute the value of π. Instead, we shall use it as the theorem leading into the heart of geometric probability, following the ideas of Crofton and Sylvester.

1.2 The space of lines

Let Graff(2,1) denote the set of all straight lines in $\mathbf{R}^2$ (the reason for this notation shall be made clearer in Chapter 7). It is well known that this set enjoys some notable properties.

To this end, denote by Z_1 the number of intersections of a straight line taken at random with a straight line segment of length L_1, and let λ_1^2 denote the invariant measure on Graff(2,1). The integral

$$\int_{\text{Graff}(2,1)} Z_1 \, d\lambda_1^2$$

depends only on L_1. Since Z_1 takes only the values 0 or 1, this integral is equal to the measure of the set of all straight lines that meet the given straight line segment. Since the value of the integral depends only on the length L_1 of the straight line segment, denote this value by $f(L_1)$. We can now repeat the argument we used for the Buffon needle problem: given a polygonal line consisting of segments of length $L_1, L_2, \ldots$, the number of intersections of a randomly chosen straight line with the polygonal line is

$$\int_{\text{Graff}(2,1)} (Z_1 + Z_2 + \cdots) \, d\lambda_1^2 = f(L_1 + L_2 + \cdots).$$

Since integrals are linear, this becomes

$$\int_{\text{Graff}(2,1)} Z_1 \, d\lambda_1^2 + \int_{\text{Graff}(2,1)} Z_2 \, d\lambda_1^2 + \cdots = f(L_1) + f(L_2) + \cdots,$$

and we again conclude that $f(L) = rL$. We shall *not* normalize the measure λ_1^2 by setting $r = 1$; rather, we shall decide later what the 'right' normalization should be.

Again we may pass to the limit. Recall that a subset K of the plane is *convex* if any two points x and y in K are the endpoints of a line segment lying inside K. A curve C in the plane is called *convex* if C encloses a convex subset. Let C be a convex curve in the plane of length L, and let

Z_C be the number of intersections of C with a randomly chosen straight line. Then

$$\int_{\text{Graff}(2,1)} Z_C \, d\lambda_1^2 = rL.$$

In particular, let K_1 and K_2 be compact convex sets in the plane with non-empty interiors, and with boundaries $C_1 = \partial K_1$ and $C_2 = \partial K_2$ of length L_1 and L_2. For each i, we have

$$\int_{\text{Graff}(2,1)} Z_{C_i} \, d\lambda_1^2 = rL_i.$$

On the other hand, since K_i is convex, a straight line meets K_i either twice or not at all (excluding the limiting cases of tangents, which can be shown to have measure zero). Thus, the function Z_{C_i} takes either the value 2 or the value 0. If we denote by D_i the set of all straight lines in $\mathbf{R}^2$ that meet K_i, then we have

$$\int_{\text{Graff}(2,1)} Z_{C_i} \, d\lambda_1^2 = 2\lambda_1^2(D_i).$$

To re-state these results in terms of probability, assume that $K_1 \subseteq K_2$. The conditional probability that a straight line shall meet the compact convex set K_1, given that it meets K_2, is the ratio

$$\frac{\lambda_1^2(D_1)}{\lambda_1^2(D_2)}.$$

The computation above shows that this ratio is equal to

$$\frac{L_1}{L_2} = \frac{\text{length}(\partial K_1)}{\text{length}(\partial K_2)}.$$

Note that the value of the normalization constant r is irrelevant to the computation of this conditional probability.

The results above (sometimes designated *Sylvester's theorem*) can be compared to the analogous result for points: if $K_1 \subseteq K_2$, the conditional probability that a point taken at random shall belong to K_1, given that it belongs to K_2, is

$$\frac{\text{area}(K_1)}{\text{area}(K_2)}.$$

Thus, we see a striking analogy: replacing every occurrence of the word 'point' by the word 'line' corresponds to replacing the word 'area' by the word 'perimeter.' This analogy suggests that a generalization of Sylvester's theorem to arbitrary dimension may prove worthwhile.

1.3 Notes

The solution to Buffon's needle problem presented here is due to Barbier [5], and was later generalized still further by Crofton in [14, 15, 16]. Crofton's main paper, which set geometric probability on its modern footing, is the Encyclopaedia Britannica article [17]. It is still an excellent reference.

In [95] Sylvester considered a variation of the Buffon needle problem in which the needle is replaced by a finite rigid collection of compact convex (and possibly disjoint) sets $K_1, \ldots, K_m$ tossed randomly into a plane tiled by evenly spaced lines. Sylvester then considered the cases in which a line meets one, some, or all of the sets K_i. In the previous section we measured the set of all lines meeting a compact convex set K in the plane. When dealing with multiple convex sets Sylvester was led to consider also the measure of the set of lines that *separate* two disjoint compact convex regions of the plane. This theme has also been pursued extensively in the work of Ambartzumian [1, 2].

Buffon's result gives a very inefficient means of approximating the number π; for a history of this technique, see [30]. For additional modern treatments of geometric probability in the plane, see also [1, 2, 49, 82, 90].

2

Valuation and integral

In Chapter 1 we expressed the Buffon needle problem in terms of a set functional (1.1) on a certain collection of sets in the plane satisfying a certain kind of additivity. We then solved the problem by characterizing this additive functional in (1.2), using in this case the fact that the functional was monotonically increasing and invariant with respect to certain motions of sets in the plane.

In this chapter we make more precise the notion of 'additive set functional', or *valuation*, on a lattice of sets. The abstract notions developed in this chapter will then be specialized to several different specific lattices in the chapters following, leading in turn to similarly elegant solutions to generalizations and analogues of Buffon's original problem. Section 2.2 is devoted to Groemer's integral theorem, which is needed to prove Groemer's extension theorems in Sections 4.1, 5.1, and 11.1.

2.1 Valuations

We now introduce a class of set functions that comprise the most basic and important tools of geometric probability, namely *valuations*. We begin with partially ordered sets and lattices. A partial ordering $\leq$ on a set L is a relation satisfying the following conditions for all $x, y, z \in L$.

(i) $x \leq x$.
(ii) If $x \leq y$ and $y \leq x$ then $x = y$.
(iii) If $x \leq y$ and $y \leq z$ then $x \leq z$.

The partially ordered set L is called a *lattice* if, for all $x, y \in L$, there exist a greatest lower bound (or *meet*) $x \wedge y \in L$ and a least upper bound (or *join*) $x \vee y \in L$. A lattice L is said to be *distributive* if, for all $x, y, z \in L$, we have the following.

(i) $x \vee (y \wedge z) = (x \vee y) \wedge (x \vee z)$.

(ii) $x \wedge (y \vee z) = (x \wedge y) \vee (x \wedge z)$.

Let S be a set, and let L be a family of subsets of S closed under finite unions and finite intersections. Such a family is clearly a distributive lattice, in which the partial ordering is given by subset inclusion, while the meet and join are given by intersection and union of sets, respectively.

A *valuation* on a lattice L of sets is a function μ defined on L that takes real values, and that satisfies the following conditions:

$$\mu(A \cup B) = \mu(A) + \mu(B) - \mu(A \cap B), \tag{2.1}$$

$$\mu(\emptyset) = 0, \text{ where } \emptyset \text{ is the empty set.} \tag{2.2}$$

By iterating the identity (2.1) we obtain the *inclusion–exclusion principle* for a valuation μ on a lattice L, namely

$$\mu(A_1 \cup A_2 \cup \cdots \cup A_n)$$
$$= \sum_i \mu(A_i) - \sum_{i<j} \mu(A_i \cap A_j) + \sum_{i<j<k} \mu(A_i \cap A_j \cap A_k) + \cdots \tag{2.3}$$

for each positive integer n.

If A is any subset of S, the *indicator function* (or simply the *indicator*) of A, denoted by I_A, is the function on S given by

- $I_A(s) = 1$; $s \in A$,
- $I_A(s) = 0$; $s \notin A$.

A finite linear combination

$$f = \sum_{i=1}^{k} \alpha_i I_{A_i}, \tag{2.4}$$

where $\alpha_i \in \mathbf{R}$, and $A_i \in L$, is said to be an *L-simple* function, or a *simple* function for short. The set of all L-simple functions forms a ring under the usual operations on functions.

Indicator functions satisfy the following properties:

$$I_{A \cap B} = I_A I_B, \tag{2.5}$$

$$I_{A \cup B} = I_A + I_B - I_A I_B = 1 - (1 - I_A)(1 - I_B). \tag{2.6}$$

By iteration of the identities (2.5) and (2.6) we obtain the inclusion–exclusion formula for indicators,

$$I_{A_1 \cup A_2 \cup \cdots \cup A_n} = 1 - (1 - I_{A_1})(1 - I_{A_2}) \cdots (1 - I_{A_n})$$

$$= \sum_i I_{A_i} - \sum_{i<j} I_{A_i \cap A_j} + \sum_{i<j<k} I_{A_i \cap A_j \cap A_k} + \cdots. \qquad (2.7)$$

A subset G of L that is closed under finite intersections is said to be a *generating set* of L when every element of L is a finite union of elements of G. Using the inclusion–exclusion formula for indicators, it can be shown that every L-simple function can be written as a finite linear combination

$$f = \sum_{i=1}^r \beta_i I_{B_i}, \qquad (2.8)$$

where $B_i \in G$. A real-valued function ν on G is called a *valuation* on G provided that ν satisfies identities (2.1) and (2.2) for all sets $A, B \in G$ such that $A \cup B \in G$ as well. Note that, since G need not be closed under unions, identity (2.1) does not make sense for all pairs of sets $A, B \in G$. Hence, there is no reason to assume that the identities (2.3) should hold for ν if $n > 2$.

Since every element $B \in L$ can be expressed as a union $B = B_1 \cup \cdots \cup B_n$ with $B_1, \ldots, B_n \in G$, we can attempt to extend ν to a valuation μ on all of L by setting

$$\mu(B) = \sum_i \nu(B_i) - \sum_{i<j} \nu(B_i \cap B_j) + \cdots, \qquad (2.9)$$

as is suggested by (2.3). There remains to check that $\mu(B)$ is well defined, in the case that B could be expressed as a union of elements of G in more than one way.

Given a valuation μ on G, define the *integral* with respect to μ as follows. For an L-simple function $f = \alpha_1 I_{A_1} + \cdots + \alpha_k I_{A_1}$, with $A_i \in G$ for $1 \le i \le k$, define

$$\int f \, d\mu = \sum_{i=1}^k \alpha_i \mu(A_i). \qquad (2.10)$$

In general, a simple function f has infinitely many expressions of the form (2.4), for A_i in G. Consequently we must check that the integral in (2.10) is well defined.

2.2 Groemer's integral theorem

The existence of the extension (2.9) and the integral (2.10) turn out to be equivalent properties of μ, a nontrivial fact stated formally as follows.

Theorem 2.2.1 (Groemer's integral theorem) *Let G be a generating set for a lattice L, and let μ be a valuation on G. The following statements are equivalent.*

(i) *μ extends uniquely to a valuation on L.*

(ii) *μ satisfies the inclusion–exclusion identities*

$$\mu(B_1 \cup B_2 \cup \cdots \cup B_n) = \sum_i \mu(B_i) - \sum_{i<j} \mu(B_i \cap B_j) + \cdots, \quad (2.11)$$

whenever $B_i \in G$ and $B_1 \cup B_2 \cup \cdots \cup B_n \in G$, and for all $n \geq 2$.

(iii) *μ defines an integral on the vector space of linear combinations of indicator functions of sets in L.*

Proof We prove the implications (i) $\Rightarrow$ (ii) $\Rightarrow$ (iii) $\Rightarrow$ (i).

If μ extends uniquely to a valuation on all of L, then (ii) follows from an iteration of identity (2.1). Therefore, (i) implies (ii).

To show that (ii) implies (iii), suppose there exist non-empty distinct $K_1, \ldots, K_m \in G$ and nonzero real numbers $\alpha_1, \ldots, \alpha_m$ such that

$$\sum_{i=1}^m \alpha_i I_{K_i} = 0, \quad (2.12)$$

while

$$\sum_{i=1}^m \alpha_i \mu(K_i) \neq 0. \quad (2.13)$$

Let $L_1 = K_1, \ldots, L_m = K_m$, $L_{m+1} = K_1 \cap K_2$, $L_{m+2} = K_1 \cap K_3$, and so on, to define a list $L_1, L_2, \ldots, L_p$, comprising all possible intersections of the sets K_i. Since G is closed under intersections, $L_i \in G$ for all i. Note also that the collection $\{L_i\}$ is closed under intersections.

Suppose that

$$\sum_{i=q}^p \alpha_i I_{L_i} = 0, \quad (2.14)$$

while

$$\sum_{i=q}^p \alpha_i \mu(L_i) \neq 0, \quad (2.15)$$

where $\alpha_q \neq 0$. Choose an instance of these equations such that q is maximal. It follows from (2.12) and (2.13) that $q \geq 1$, while the conditions (2.14) and (2.15) imply that $q < p$.

Suppose that $x \in L_q - \bigcup_{j=q+1}^{p} L_j$. Then (2.14) implies that

$$\alpha_q = \sum_{i=q}^{p} \alpha_i I_{L_i}(x) = 0,$$

contradicting our assumption. It follows that

$$L_q \subseteq L_{q+1} \cup \cdots \cup L_p,$$

so that

$$L_q = L_q \cap (L_{q+1} \cup \cdots \cup L_p) = (L_q \cap L_{q+1}) \cup \cdots \cup (L_q \cap L_p).$$

For $i > q$, note that $L_q \cap L_i = L_j$, where $j > q$. Using the principle of inclusion–exclusion **(ii)** we obtain

$$\sum_{i=q}^{p} \alpha_i \mu(L_i) = \alpha_q \mu \left(\bigcup_{i=q+1}^{p} (L_q \cap L_i) \right) + \sum_{i=q+1}^{p} \alpha_i \mu(L_i) = \sum_{i=q+1}^{p} \beta_i \mu(L_i),$$

so that

$$\sum_{i=q+1}^{p} \beta_i \mu(L_i) \neq 0 \qquad (2.16)$$

by (2.15), where each β_i is obtained by collecting the terms containing $\mu(L_i)$. Meanwhile, application of the same inclusion–exclusion procedure to the indicator functions yields

$$\sum_{i=q}^{p} \alpha_i I_{L_i} = \alpha_q I_{\bigcup_{i=q+1}^{p}(L_q \cap L_i)} + \sum_{i=q+1}^{p} \alpha_i I_{L_i} = \sum_{i=q+1}^{p} \beta_i I_{L_i}$$

so that

$$\sum_{i=q+1}^{p} \beta_i I_{L_i} = 0 \qquad (2.17)$$

by (2.14). Together (2.16) and (2.17) contradict the maximality of q. This completes the proof that **(ii)** implies **(iii)**.

To show that **(iii)** implies **(i)**, suppose that the function μ defines an integral on the space of L-simple functions. For $A \in L$ define

$$\mu(A) = \int I_A \, d\mu.$$

The linearity of the integral together with the identity (2.6) implies that this extension of μ is a valuation on L. □

A linear functional T on the vector space of simple functions determines a valuation μ by setting

$$\mu(A) = T(I_A)$$

for every $A \in L$. It is easily verified that, for a simple function f,

$$T(f) = \int f \, d\mu.$$

Thus, insofar as simple functions are concerned, there is a bijective correspondence between linear functionals and valuations.

Let $B(L)$ be the relative Boolean algebra generated by the distributive lattice L; that is, the smallest family of subsets of S containing L that is closed under finite unions, finite intersections, and relative complements. Note that for $A, B \in L$

$$I_{A-B} = I_{A-(A\cap B)} = I_A - I_A I_B. \tag{2.18}$$

Let $\mathcal{I}(L)$ denote the algebra of simple functions generated by finite sums, products, and differences of indicator functions of sets in L. If follows from (2.5), (2.6), and (2.18) that $I_C \in \mathcal{I}(L)$ for all $C \in B(L)$.

Corollary 2.2.2 *A valuation μ defined on a distributive lattice L has a uniquely defined extension to the Boolean algebra $B(L)$.*

Proof By Theorem 2.2.1, μ defines an integral on the space of indicator functions $\mathcal{I}(L)$. For $C \in B(L)$ define

$$\mu(C) = \int I_C \, d\mu.$$

The linearity of the integral together with identity (2.7) implies that this extension of μ is a valuation on $B(L)$. □

2.3 Notes

The study of valuations, while natural enough on its own as a finitely additive precursor to the measure theory of modern probability, was invigorated especially by interest in dissection problems on polytopes, and Hilbert's third problem in particular [8, 71, 72, 81] (see also Section 8.6). In the sections that follow we shall see that most of the interesting functionals of geometric probability satisfy the valuation property in some respect.

The integral and extension theorems of Groemer may be found in [32]. McMullen and Schneider gave a thorough survey of the modern theory of valuations on convex bodies in [72], later updated by McMullen in [71].

3

A discrete lattice

In this chapter we focus on combinatorial properties of the lattice of subsets of a finite set, properties which carry over in analogous forms to the lattice of parallelotopes in Chapter 4, of subspaces in Chapter 6, of polyconvex sets in Chapters 5, 7–10, and of spherical polyconvex sets in Chapter 11. An especially important result of this chapter is the characterization of valuations invariant under the permutation group. The idea of characterizing valuations invariant with respect to a group action or a set of symmetries is central to our treatment of geometric probability; this theme will recur frequently in the chapters following.

3.1 Subsets of a finite set

Let S be a non-empty set with n elements, and denote by $P(S)$ the set of all subsets of S, partially ordered by subset inclusion. The set $P(S)$ is a (finite) Boolean algebra of subsets.

Recall that the union and intersection of sets coincide with the least upper bound and greatest lower bound in the partially ordered set $P(S)$. We denote the elements of $P(S)$ by lower case letters x, y, etc.

A *segment* of $P(S)$, denoted by $[x, y]$, where $x \leq y$, consists of all elements $z \in P(S)$ such that $x \leq z \leq y$. Every segment $[x, y]$ is naturally isomorphic to the Boolean algebra $P(y - x)$.

A *chain* in $P(S)$ is a linearly ordered subset; that is, a subset in which, for every pair x, y, either $x \leq y$ or $y \leq x$. An *antichain* is a subset $A \subseteq P(S)$ such that, if $x, y \in A$, then neither $x < y$ nor $y < x$. A *flag* F in $P(S)$ is a maximal chain; that is, a chain such that if $G \supseteq F$ and G is a chain, then $G = F$. These notions apply to all partially ordered sets.

13

For $x \in P(S)$, the *rank* $r(x)$ is the number of elements of the set x. The antichain consisting of all elements of $P(S)$ of rank k shall be denoted by $P_k(S)$. The *size*, or number of elements, of $P_k(S)$ is the binomial coefficient

$$\binom{n}{k}.$$

A flag in $P(S)$ is naturally identified with a linear order $(s_1, s_2, \ldots, s_n)$ on the one-element subsets s_i of S. Hence, there are $n!$ flags in $P(S)$. A flag contains an element $x \in P_k(S)$ whenever $x = s_1 \cup s_2 \cup \cdots \cup s_k$. Thus, there are $k!(n-k)!$ flags containing a particular $x \in P_k(S)$. This elementary argument gives the classical expression for the binomial coefficient:

$$\binom{n}{k} = \frac{n!}{k!(n-k)!}.$$

Actually, the same argument can be made to yield a much stronger result. Denote by $|A|$ the size of a finite set A.

Theorem 3.1.1 (Sperner's theorem) *Let A be an antichain in $P(S)$. Then*

$$|A| \le \binom{n}{\langle n/2 \rangle}.$$

Here the expression $\langle n/2 \rangle$ denotes the greatest integer smaller than or equal to $n/2$. Evidently equality is attained in Theorem 3.1.1 when $A = P_{\langle n/2 \rangle}(S)$.

Proof The proof of this theorem depends on a more precise result, known as the Lubell–Yamamoto–Meshalkin (L.Y.M.) inequality. Let A be an antichain in $P(S)$, and let A_k consist of all elements of A of rank k. Then

$$\sum_{k=0}^{n} \frac{|A_k|}{\binom{n}{k}} \le 1. \qquad (3.1)$$

To prove (3.1), notice that every flag meets A in at most one element of $P(S)$. Therefore, the number p of flags meeting A is

$$p = \sum_{k=0}^{n} k!(n-k)!|A_k|.$$

Since there are $n!$ flags in $P(S)$, we have $p \leq n!$. Dividing by $n!$ proves the L.Y.M. inequality (3.1).

To complete the proof of Sperner's theorem, recall that

$$\binom{n}{k} \leq \binom{n}{\langle n/2 \rangle},$$

for all $0 \leq k \leq n$. The L.Y.M. inequality (3.1) now gives

$$\sum_{k=0}^{n} \frac{|A_k|}{\binom{n}{\langle n/2 \rangle}} \leq \sum_{k=0}^{n} \frac{|A_k|}{\binom{n}{k}} \leq 1;$$

that is,

$$|A| = \sum_{k=0}^{n} |A_k| \leq \binom{n}{\langle n/2 \rangle}.$$

$\square$

Suppose that $1 \leq r \leq n+1$ is a integer. A subset $F \subseteq P(S)$ is called an r-*family* if chains in F contain no more than r elements. For example, an antichain is a 1-family. Given an r-family F in $P(S)$ let $F_k = F \cap P_k(S)$. Since every flag in $P(S)$ meets F in at most r elements, we have

$$\sum_{k=0}^{n} k!(n-k)!|F_k| \leq n! \cdot r.$$

We then obtain the following generalization of (3.1):

$$\sum_{k=0}^{n} \frac{|F_k|}{\binom{n}{k}} \leq r. \qquad (3.2)$$

This inequality leads in turn to a generalization of Sperner's theorem to r-families:

Theorem 3.1.2 *Let F be an r-family in $P(S)$. Then*

$$|F| \leq \binom{n}{\langle \frac{n+1}{2} \rangle} + \binom{n}{\langle \frac{n+2}{2} \rangle} + \cdots + \binom{n}{\langle \frac{n+r}{2} \rangle}.$$

In order to prove Theorem 3.1.2 we make use of the following lemma.

Lemma 3.1.3 *Suppose that $c_0 \geq c_1 \geq \cdots \geq c_n > 0$. If $c_i \geq x_i \geq 0$ for $0 \leq i \leq n$, and if*

$$x_0 + x_1 + \cdots + x_n \geq c_0 + c_1 + \cdots + c_{r-1}, \qquad (3.3)$$

then

$$\sum_{k=0}^{n} \frac{x_k}{c_k} \geq r. \qquad (3.4)$$

If $c_0 > \cdots > c_n > 0$, then equality holds in (3.4) if and only if $x_i = c_i$ for $0 \leq i \leq r-1$ and $x_i = 0$ for $r \leq i \leq n$.

Proof Suppose that $x_0, \ldots, x_n \geq 0$ minimize the sum

$$\sum_{k=0}^{n} \frac{x_k}{c_k}, \qquad (3.5)$$

subject to the condition (3.3). If this minimum value of (3.5) is less than r then $x_i < c_i$ for some $i \leq r-1$. Let i be the smallest index such that $x_i < c_i$, so that $x_k/c_k = 1$ for all $0 \leq k < i$ (unless $i = 0$).

The inequality (3.3) then implies that there is an index $j \geq r$ such that $x_j > 0$. Let j be the largest such index, so that $x_k = 0$ for $k > j$ (unless $j = n$). In particular, note that $j > i$.

Suppose $c_i = c_j$. Since $c_i \geq c_{i+1} \cdots \geq c_j$, we have $c_i = c_{i+1} = \cdots = c_j$. It then follows that

$$
\begin{aligned}
c_0 + \cdots + c_{i-1} + x_i + \cdots + x_j &= x_0 + \cdots + x_n \\
&\geq c_0 + \cdots + c_{r-1} \\
&= c_0 + \cdots + c_{i-1} + (r-i)c_i,
\end{aligned}
$$

since $r - 1 < j$. Therefore,

$$x_i + \cdots + x_j \geq (r-i)c_i. \qquad (3.6)$$

It follows that

$$
\begin{aligned}
\sum_{k=0}^{n} \frac{x_k}{c_k} &= \sum_{k=0}^{i-1} \frac{x_k}{c_k} + \sum_{k=i}^{j} \frac{x_k}{c_k} \\
&= i + \sum_{k=i}^{j} \frac{x_k}{c_k} \\
&= i + \frac{1}{c_i}(x_i + \cdots + x_j) \\
&\geq i + (r-i)\frac{c_i}{c_i},
\end{aligned}
$$

by (3.6). In other words,

$$\sum_{k=0}^{n} \frac{x_k}{c_k} \geq r. \tag{3.7}$$

This contradicts the assumption that $x_0, \ldots, x_n$ minimizes (3.5) at a value less than r. Therefore, inequality (3.4) follows.

Now suppose instead that $c_i > c_j$. If $x_i + x_j \leq c_i$ set $y_i = x_i + x_j$ and $y_j = 0$. Otherwise, set $y_i = c_i$ and $y_j = x_j - (c_i - x_i)$. In either case set all other $y_k = x_k$.

Since $c_i > c_j$ it follows in both cases that

$$\frac{y_i}{c_i} + \frac{y_j}{c_j} < \frac{x_i}{c_i} + \frac{x_j}{c_j},$$

so that

$$\sum_{k=0}^{n} \frac{y_k}{c_k} < \sum_{k=0}^{n} \frac{x_k}{c_k},$$

contradicting the assumption that $x_0, \ldots, x_n$ minimizes (3.5). Inequality (3.4) now follows again for this case.

If $c_0 > \cdots > c_n$, then $c_i > c_j$ is guaranteed, and so it follows that (3.5) is minimized only if $x_i = c_i$ for $0 \leq i \leq r - 1$ and $x_i = 0$ for $r \leq i \leq n$. □

Proof of Theorem 3.1.2 Relabel the binomial coefficients $c_0, c_1, \ldots, c_n$ in *descending* order, so that $c_0 \geq c_1 \geq \cdots \geq c_n > 0$; then relabel the numerators $|F_k|$ in (3.2) by $x_0, x_1, \ldots, x_n$, so that each x_k is the numerator of that term of (3.2) having c_k as denominator. The inequality (3.2) now becomes

$$\sum_{k=0}^{n} \frac{x_k}{c_k} \leq r.$$

It then follows from Lemma 3.1.3 that

$$x_0 + x_1 + \cdots + x_n \leq c_0 + c_1 + \cdots + c_{r-1}.$$

In other words,

$$|F| = \sum_{k=0}^{n} |F_k| \leq \binom{n}{\langle \frac{n+1}{2} \rangle} + \binom{n}{\langle \frac{n+2}{2} \rangle} + \cdots + \binom{n}{\langle \frac{n+r}{2} \rangle}.$$

□

The theory of binomial coefficients and antichains generalizes nicely to a theory of multinomial coefficients and special collections of ordered partitions, known as s-systems. Sperner's result can also be generalized from a bound on the size of an antichain to a bound on the size of an s-system.

A map $\delta : \{1, \ldots, r\} \longrightarrow P(S)$ is called an *r-decomposition of S* if

(i) $\delta(i) \cap \delta(j) = \emptyset$ for $i \neq j$, and
(ii) $\delta(1) \cup \cdots \cup \delta(r) = S$.

Denote by $\mathrm{Dec}(S, r)$ the set of all *r*-decompositions of S. Note that for each $\delta \in \mathrm{Dec}(S, r)$

$$|\delta(1)| + \cdots + |\delta(r)| = n.$$

Given non-negative integers $a_1, a_2, \ldots, a_r$ such that $a_1 + \cdots + a_r = n$ we denote by $P_{a_1, \ldots, a_r}(S)$ the set of all *r*-decompositions δ such that $|\delta(i)| = a_i$ for $i = 1, \ldots, r$. In other words, $P_{a_1, \ldots, a_r}(S)$ is the set of all (ordered) partitions of S into disjoint unions of subsets having sizes $a_1, \ldots, a_r$. Evidently the set $\mathrm{Dec}(S, r)$ can be expressed as the finite disjoint union

$$\mathrm{Dec}(S, r) = \biguplus_{a_1 + \cdots + a_r = n} P_{a_1, \ldots, a_r}(S).$$

The size of $P_{a_1, \ldots, a_r}(S)$ is given by the multinomial coefficient

$$|P_{a_1, \ldots, a_r}(S)| = \binom{n}{a_1, \ldots, a_r}.$$

An *s-system* of order r (or an *s-system in* $\mathrm{Dec}(S, r)$) is a subset $\sigma \subseteq \mathrm{Dec}(S, r)$ such that the set

$$\{\delta(i) : \delta \in \sigma\} \tag{3.8}$$

is an antichain in $P(S)$ for each $1 \leq i \leq r$.

An obvious example of an s-system of order r is $P_{a_1, \ldots, a_r}(S)$ for some admissible selection of $a_1, \ldots, a_r$. If $\delta, \zeta \in P_{a_1, \ldots, a_r}(S)$ then $\delta(i)$ and $\zeta(i)$ both have size a_i, so that either $\delta(i) = \zeta(i)$ or the two sets are incomparable in the subset partial ordering on $P(S)$. This holds for $i = 1, \ldots, r$, and so the antichain condition on (3.8) is satisfied.

Other disguised examples with which we have already worked are the s-systems of order 2. Let A be an antichain in $P(S)$. For each $x \in A$ we can express S as the disjoint union $x \uplus S - x$, so that the pair $(x, S - x)$ is a 2-decomposition in $\mathrm{Dec}(S, 2)$. Moreover, the set

$$\{S - x : x \in A\}$$

is also an antichain in $P(S)$, so that the set

$$\sigma = \{(x, S - x) : x \in A\}$$

is an s-system of order 2. Thus the notion of s-system is a generalization of the notion of an antichain. Similarly, the collection $P_k(S)$ can also be viewed as $P_{k,n-k}(S)$ through the bijection $x \mapsto (x, S - x)$.

For $\delta \in P_{a_1,\dots,a_r}(S)$ we say that a flag $(x_0, x_1, \dots, x_n)$ in $P(S)$ is *compatible* with δ if

(i) $x_{a_1} = \delta(1)$, and
(ii) $x_{a_1 + \dots + a_i} - x_{a_1 + \dots + a_{i-1}} = \delta(i)$, for $i \geq 2$.

Here the difference $x_{a_1 + \dots + a_i} - x_{a_1 + \dots + a_{i-1}}$ denotes the complement of the set $x_{a_1 + \dots + a_{i-1}}$ inside the larger set $x_{a_1 + \dots + a_i}$. For $A \subseteq P_{a_1,\dots,a_r}$, let $\mathrm{Flag}(A)$ be the set of all flags $(x_0, x_1, \dots, x_n)$ compatible with some $\delta \in A$.

Note that for each $\delta \in P_{a_1,\dots,a_r}(S)$, there are exactly $a_1! a_2! \cdots a_r!$ flags compatible with δ; that is, to choose a flag compatible with δ one must choose a permutation of the a_1 elements of $\delta(1)$, of which there are $a_1!$, and then a permutation of the a_2 elements of $\delta(2)$, of which there are $a_2!$, and so on up through $\delta(r)$. Since each of the $n!$ flags is compatible with one (and only one) r-decomposition in $P_{a_1,\dots,a_r}(S)$, it follows that

$$\binom{n}{a_1, a_2, \dots, a_r} = \frac{n!}{a_1! a_2! \cdots a_r!}.$$

Just as Sperner's Theorem 3.1.1 gives the maximum possible measure for an antichain A in $P(S)$, a generalization of this theorem gives the maximum possible size for an s-system in $\mathrm{Dec}(S, r)$. *En route* to such a generalization we prove a multinomial version of the L.Y.M. inequality.

Theorem 3.1.4 (The multinomial L.Y.M. inequality) *Let $\sigma \subseteq \mathrm{Dec}(S, r)$ be an s-system. For $a_1 + \cdots + a_r = n$ let*

$$\sigma_{a_1,\dots,a_r} = \sigma \cap P_{a_1,\dots,a_r}(S),$$

so that

$$\sigma = \bigcup_{a_1 + \cdots + a_r = n} \sigma_{a_1,\dots,a_r}$$

is a disjoint union. Then

$$\sum_{a_1 + \cdots + a_r = n} \frac{|\sigma_{a_1,\dots,a_r}|}{\binom{n}{a_1,\dots,a_r}} \leq 1. \tag{3.9}$$

Proof For $a_1 + \cdots + a_r = n$ the number of flags compatible with $\sigma_{a_1,\ldots,a_r}$ is given by

$$|\text{Flag}(\sigma_{a_1,\ldots,a_r})| = |\sigma_{a_1,\ldots,a_r}| a_1! \cdots a_r!$$

Suppose that a flag $(x_0, x_1, \ldots, x_n)$ is compatible with both $\gamma, \delta \in \sigma$. Then $\gamma(1) = x_{a_1}$ and $\delta(1) = x_{b_1}$, where $a_1 = |\gamma(1)|$ and $b_1 = |\delta(1)|$. Since $(x_0, x_1, \ldots, x_n)$ is a flag, we have $x_{a_1} \subseteq x_{b_1}$ or vice versa. However, σ is an s-system, so that either $\gamma(1) = \delta(1)$ or the two sets are incomparable. Therefore $\gamma(1) = \delta(1)$ and $a_1 = b_1$. Continuing, we have $\gamma(2) = x_{a_1+a_2} - x_{a_1}$ and $\delta(2) = x_{b_1+b_2} - x_{a_1}$ (since $a_1 = b_1$). A similar argument then implies that $\gamma(2) = \delta(2)$ and $a_2 = b_2$. Continuing in this manner we conclude that $\gamma(i) = \delta(i)$ for each $1 \le i \le r$, so that $\gamma = \delta$. In other words, every flag in $P(S)$ is compatible with at most one r-decomposition $\delta \in \sigma$. It follows that

$$\sum_{a_1+\cdots+a_r=n} |\sigma_{a_1,\ldots,a_r}| a_1! \cdots a_r!$$

$$= \sum_{a_1+\cdots+a_r=n} |\text{Flag}(\sigma_{a_1,\ldots,a_r})| = |\text{Flag}(\sigma)| \le n!$$

so that

$$\sum_{a_1+\cdots+a_r=n} \frac{|\sigma_{a_1,\ldots,a_r}|}{\binom{n}{a_1,\ldots,a_r}} \le 1.$$

$\square$

We are now ready to prove a multinomial generalization of Sperner's Theorem 3.1.1.

Theorem 3.1.5 (Meshalkin's theorem) *Let σ be an s-system in* $\text{Dec}(S, r)$. *Then*

$$|\sigma| \le \binom{n}{\underbrace{\langle n/r \rangle, \cdots, \langle n/r \rangle}_{r-b}, \underbrace{\langle n/r \rangle + 1, \cdots, \langle n/r \rangle + 1}_{b}} \qquad (3.10)$$

where $n \equiv b \bmod r$.

Here $\langle n/r \rangle$ denotes the largest positive integer less than or equal to n/r.

Proof It is not difficult to show that, for all compositions $a_1 + \cdots + a_r = n$ of a positive integer n, we have

$$\binom{n}{a_1, \cdots, a_r} \le \binom{n}{\langle n/r \rangle, \cdots, \langle n/r \rangle, \langle n/r \rangle + 1, \cdots, \langle n/r \rangle + 1}.$$

For a sketch of a proof, see the analogous Proposition 6.5.2.

Let $\sigma_{a_1,\ldots,a_r} = \sigma \cap P_{a_1,\ldots,a_r}(S)$. It now follows from (3.9) that

$$\sum_{a_1+\cdots+a_r=n} \frac{|\sigma_{a_1,\ldots,a_r}|}{\left(\genfrac{}{}{0pt}{}{n}{\langle n/r\rangle,\cdots,\langle n/r\rangle,\langle n/r\rangle+1,\cdots,\langle n/r\rangle+1}\right)} \leq \sum_{a_1+\cdots+a_r=n} \frac{|\sigma_{a_1,\ldots,a_r}|}{\left(\genfrac{}{}{0pt}{}{n}{a_1,\ldots,a_r}\right)} \leq 1,$$

so that

$$|\sigma|$$
$$= \sum_{a_1+\cdots+a_r=n} |\sigma_{a_1,\ldots,a_r}| \leq \left(\genfrac{}{}{0pt}{}{n}{\langle n/r\rangle,\cdots,\langle n/r\rangle,\langle n/r\rangle+1,\cdots,\langle n/r\rangle+1}\right).$$

$\square$

Several other results of extremal set theory follow from the L.Y.M. inequality or its variants, but we reluctantly move on to the next topic.

3.2 Valuations on a simplicial complex

Define a *simplicial complex* to be a subset A of $P(S)$ such that if $x \in A$ and $y \leq x$ then $y \in A$. A simplicial complex is a partially ordered set in the order induced by $P(S)$. The set of maximal elements of a simplicial complex is an antichain. A simplicial complex having exactly one maximal element is called a *simplex*. A simplex whose unique maximal element is a set of size k is called a *k-simplex*.

The (set-theoretic) union and intersection of any number of simplicial complexes is again a simplicial complex. Thus, the set $L(S)$ of all simplicial complexes in $P(S)$ is a distributive lattice, and we can study valuations on $L(S)$.

For $x \in P(S)$, denote by $\bar{x}$ the simplex whose maximal element is x; that is, the set of all $y \in P(S)$ such that $y \leq x$.

It follows from Groemer's integral Theorem 2.2.1 and Corollary 2.2.2 that every valuation μ on $L(S)$ extends uniquely to a valuation, again denoted by μ, on the Boolean algebra $P(P(S))$ of all subsets of $P(S)$, which is generated by $L(S)$. Such a valuation is evidently determined by its value on the one-element subsets of $P(S)$; that is, by arbitrarily assigning a value $\mu(\{x\})$ for each $x \in P(S)$.

Let x be of rank k, and let $A_1, A_2, \ldots, A_k$ be the maximal simplices $A_i \in \bar{x}$ such that $A_i \neq \bar{x}$. (These simplices are sometimes called the *facets* of $\bar{x}$.) Then

$$\mu(\{x\}) = \mu(\bar{x}) - \mu(A_1 \cup A_2 \cup \cdots \cup A_k).$$

The right-hand side can be computed in terms of simplices of lower rank, by the inclusion–exclusion principle. Thus, by induction on the rank, we have the following theorem.

Theorem 3.2.1 *Every valuation μ on the distributive lattice $L(S)$ of all simplicial complexes is uniquely determined by the values $\mu(\overline{x})$, $x \in P(S)$. The values $\mu(\overline{x})$ may be arbitrarily assigned.*   □

A valuation μ on $L(S)$ is called *invariant* if it is invariant under the group of permutations of the set S; that is, if $\mu(A) = \mu(gA)$ for every simplicial complex A and for every permutation g of the set S (which induces a permutation on $L(S)$, also denoted by g). We next establish the existence of the Euler characteristic. The following is an immediate consequence of Theorem 3.2.1.

Theorem 3.2.2 (The existence of the Euler characteristic) *There exists a unique invariant valuation μ on $L(S)$, called the Euler characteristic, such that $\mu_0(\overline{x}) = 1$ for every simplex $\overline{x}$ with $r(x) > 0$, and such that $\mu_0(\overline{\emptyset}) = 0$.*   □

Next, we derive the classical alternating formula for the Euler characteristic. Define a valuation on $P(P(S))$, denoted μ'_0, by setting

$$\mu'_0(\overline{\emptyset}) = 0,$$

and

$$\mu'_0(\{x\}) = (-1)^{k-1},$$

if $r(x) = k$. Then

$$\mu'_0(\overline{x}) = \sum_{y \leq x} \mu'_0(\{y\})$$

$$= \sum_{y \leq x,\, r(y)=1} \mu'_0(\{y\}) + \sum_{y \leq x,\, r(y)=2} \mu'_0(\{y\}) + \cdots + \mu'_0(\{x\}).$$

Since the simplex $\overline{x}$ contains

$$\binom{k}{j}$$

elements of rank j, the right-hand side simplifies to

$$\binom{k}{1} - \binom{k}{2} + \binom{k}{3} - \cdots + (-1)^{k+1}\binom{k}{k} = 1,$$

so that $\mu'_0(\overline{x}) = \mu_0(\overline{x})$, for all simplices $\overline{x}$. It now follows from Theorem 3.2.1 that $\mu'_0 = \mu_0$, and that the following formula holds.

Theorem 3.2.3 (The discrete Euler formula) *Let A be a simplicial complex, and let f_k be the number of elements (or 'faces') of rank k. Then*

$$\mu_0(A) = f_1 - f_2 + f_3 - \cdots. \tag{3.11}$$

$\square$

For $i > 0$, set

$$\mu_i(\overline{x}) = |\overline{x} \cap P_i(S)|,$$

and extend μ_i to all of $L(S)$ by Theorem 3.2.1. Clearly, for every simplicial complex A,

$$\mu_i(A) = |A \cap P_i(S)|.$$

The discrete Euler formula can now be rewritten

$$\mu_0(A) = \mu_1(A) - \mu_2(A) + \mu_3(A) - \cdots, \tag{3.12}$$

for any simplicial complex A.

The valuations μ_k can also be expressed in terms of symmetric functions. Let $P_1(S) = \{a_1, a_2, \ldots, a_n\}$. Given the symmetric function

$$e_k(t_1, t_2, \ldots, t_n) = \sum_{1 \leq i_1 < \cdots < i_k \leq n} t_{i_1} t_{i_2} \ldots t_{i_k},$$

let $t_i(\overline{x}) = 1$ if $a_i \in \overline{x}$ and $t_i(\overline{x}) = 0$ if $a_i \notin \overline{x}$. Evaluated at $\overline{x}$, for $x \neq \emptyset$, we have $t_{i_1} t_{i_2} \ldots t_{i_k} = 1$ if $\{a_{i_1}, \ldots, a_{i_k}\} \in \overline{x}$, and $t_{i_1} t_{i_2} \ldots t_{i_k} = 0$ otherwise. It follows that $e_k(t_1, t_2, \ldots, t_n) = \mu_k(\overline{x})$ for every simplex $\overline{x}$ other than $\{\emptyset\}$.

Note also that if $x \in P(S)$ and $r(x) = j$ then $\mu_i(\{x\}) = 1$ if $i = j$, while $\mu_i(\{x\}) = 0$ if $i \neq j$.

Theorem 3.2.4 (The discrete basis theorem) *The invariant valuations $\mu_0, \mu_1, \ldots, \mu_n$ span the vector space of all invariant valuations μ on $L(S)$ such that $\mu(\{\emptyset\}) = 0$. The only linear relation among them is formula (3.12).*

Proof Suppose μ is an invariant valuation on $L(S)$ such that $\mu(\{\emptyset\}) = 0$. Extend μ to all of $P(P(S))$. Note that the extended valuation, which is still denoted μ, is again invariant. If x and y have the same rank in $P(S)$, say $r(x) = r(y) = i$, then there exists a permutation g of S such that $gx = y$. Therefore, $\mu(\{x\}) = \mu(\{y\}) = c_i$, for some constant c_i. Thus, the valuation

$$\mu - \sum_{i=1}^{n} c_i \mu_i$$

vanishes on all singleton sets $\{x\}$, for all $x \in P(S)$, and therefore vanishes on all of $P(P(S))$.   □

As an application of the discrete basis theorem, we shall derive a discrete analogue of the kinematic formula (whose classical geometric version appears in Chapter 10).

One way to construct invariant valuations on $L(S)$ is the following. Start with any valuation μ on $L(S)$ such that $\mu(\{\emptyset\}) = 0$, and let B be any simplicial complex. For any simplicial complex A, set

$$\mu(A; B) = \frac{1}{n!} \sum_{g} \mu(A \cap gB),$$

where g ranges over all permutations of the set S of size n. For fixed A, the set function $\mu(A; B)$ is a valuation in the variable B; in fact, it is an invariant valuation. It can therefore be expressed as a linear combination of the valuations μ_i, with coefficients $c_i(A)$ depending on A:

$$\mu(A; B) = \sum_{i=1}^{n} c_i(A)\mu_i(B). \tag{3.13}$$

Meanwhile, for fixed B, the set function $\mu(A; B)$ is a valuation in the variable A. From this it follows that each of the coefficients $c_i(A)$ is a valuation in the variable A. The coefficient $c_i(A)$ can be given an explicit expression in terms of μ. This follows from the fact that if $x \in P(S)$ and $r(x) = j$ then $\mu_i(\{x\}) = 1$ if $i = j$ and is zero if $i \neq j$.

Consider the case in which μ is an *invariant* valuation. If so, then

$$\mu(A; B) \;=\; \frac{1}{n!} \sum_{g} \mu(A \cap gB) = \frac{1}{n!} \sum_{g} \mu(g^{-1}A \cap B)$$

$$\;=\; \frac{1}{n!} \sum_{g} \mu(gA \cap B) = \mu(B; A).$$

Moreover, the coefficients $c_i(A)$ are now invariant valuations in the variable A. Therefore, Theorem 3.2.4 implies that

$$\mu(A; B) = \sum_{i,j=1}^{n} c_{ij}\mu_i(A)\mu_j(B).$$

Since $\mu(A;B) = \mu(B;A)$, it is evident that $c_{ij} = c_{ji}$. It turns out that most of the constants c_{ij} are equal to zero. In order to compute the constants c_{ij} explicitly, extend the valuation μ to the Boolean algebra $P(P(S))$ generated by $L(S)$, and let α_i denote the value of μ on a singleton set in $P(P(S))$ whose element is a subset of S of size i.

Theorem 3.2.5 (The discrete kinematic formula) *Suppose that μ is an invariant valuation on $L(S)$. For all $A, B \in L(S)$,*

$$\mu(A;B) = \sum_{i=1}^{n} \binom{n}{i}^{-1} \alpha_i \mu_i(A) \mu_i(B).$$

Proof Suppose that $x_i, y_j \subset S$ with size i and j respectively. Let $A = \{x_i\}$ and $B = \{y_j\}$. For any permutation g of S, the set $A \cap gB = \emptyset$ if $i \neq j$. If $i = j$ then $A \cap gB = \emptyset$ if $x_i \neq gy_j$. Since there are $i!(n-i)!$ permutations g of S such that $x_i = gy_j$ (if $i = j$), we have

$$\mu(A;B) = \frac{1}{n!} \sum_{g} \mu(A \cap gB) = \frac{i!(n-i)!}{n!} \mu(A) = \binom{n}{i}^{-1} \alpha_i.$$

Meanwhile, $\mu_k(A) = 1$ if $k = i$ and is equal to zero otherwise. Similarly, $\mu_k(B) = 1$ if $k = j$ and is equal to zero otherwise. Hence,

$$\mu(A;B) = \sum_{i,j=1}^{n} c_{ij} \mu_i(A) \mu_j(B) = c_{ij}.$$

Therefore,

$$c_{ij} = \binom{n}{i}^{-1} \alpha_i$$

if $i = j$ and is equal to zero otherwise. $\square$

We shall be particularly concerned with the case $\mu = \mu_0$. The discrete Euler formula (3.11) implies that $\mu_0(\{x_i\}) = (-1)^{i+1}$, so that

$$\frac{1}{n!} \sum_{g} \mu_0(A \cap gB) = \sum_{i=1}^{n} (-1)^{i+1} \binom{n}{i}^{-1} \mu_i(A) \mu_i(B),$$

for all $A, B \in L(S)$.

If $\overline{x}$ and $\overline{y}$ are simplices, then either $\overline{x} \cap \overline{y}$ is a smaller (non-empty) simplex, in which case $\mu_0(\overline{x} \cap \overline{y}) = 1$, or $\overline{x} \cap \overline{y} = \emptyset$, in which case

$\mu_0(\overline{x} \cap \overline{y}) = 0$. The probability that a randomly chosen k-simplex $\overline{x}_k$ shall meet a fixed l-simplex $\overline{y}_l$ can now be computed as follows:

$$\frac{1}{n!} \sum_g \mu_0(\overline{y}_l \cap g\overline{x}_k) = \sum_{i=1}^{n} (-1)^{i+1} \binom{n}{i}^{-1} \mu_i(\overline{y}_l)\mu_i(\overline{x}_k)$$

$$= \sum_{i=1}^{n} (-1)^{i+1} \binom{n}{i}^{-1} \frac{l!}{i!(l-i)!} \frac{k!}{i!(k-i)!},$$

so that

$$\frac{1}{n!} \sum_g \mu_0(\overline{y}_l \cap g\overline{x}_k) = \sum_{i=1}^{n} (-1)^{i+1} \binom{n}{i}^{-1} \binom{l}{i}\binom{k}{i}. \tag{3.14}$$

The equation (3.14) leads to a notable example of how the discrete kinematic formula can be used to generate identities for the binomial coefficients. To generate such an identity, we use elementary probabilistic reasoning to compute instead the probability that $y_l \cap gx_k = \emptyset$, for a random permutation g. Label the elements of S by $\{s_1, \ldots, s_n\}$ so that $x_k = \{s_1, \ldots, s_k\}$. In order for $y_l \cap gx_k = \emptyset$ to hold, we require $gs_1 \in S - y_l$, of which there are $n-l$ choices. There then remain $n-(l+1)$ possible values for gs_2, etc., so that there are

$$(n-l)(n-l-1)\cdots(n-l-k+1)$$

choices of values for $gs_1, \ldots, gs_k$. Having chosen these values, there are $n-k$ possible choices remaining for gs_{k+1}, then $n-k-1$ possible values for gs_{k+2}, and so on, up to one possible value remaining for gs_n. It follows that there are

$$(n-l)\cdots(n-l-k+1)(n-k)\cdots 1 = \frac{(n-l)!(n-k)!}{(n-k-l)!}$$

permutations g of S such that $y_l \cap gx_k = \emptyset$. Hence the probability that $y_l \cap gx_k = \emptyset$ for a random permutation g is given by

$$\frac{1}{n!}\frac{(n-l)!(n-k)!}{(n-k-l)!} = \frac{k!(n-k)!(n-l)!}{n!k!(n-k-l)!} = \binom{n}{k}^{-1}\binom{n-l}{k}.$$

It now follows from (3.14) that

$$\sum_{i=1}^{n} (-1)^{i+1} \binom{n}{i}^{-1} \binom{l}{i}\binom{k}{i} = 1 - \binom{n}{k}^{-1}\binom{n-l}{k}. \tag{3.15}$$

By adding the term corresponding to $i=0$ to both sides of (3.15) and multiplying by -1 we obtain the following identity:

Theorem 3.2.6

$$\sum_{i=0}^{n}(-1)^i\binom{n}{i}^{-1}\binom{l}{i}\binom{k}{i} = \binom{n}{k}^{-1}\binom{n-l}{k},\qquad(3.16)$$

for all positive integers $0 \le k, l \le n$.   □

Note that, if $k + l > n$, then

$$\binom{n-l}{k} = 0.$$

In the preceding argument this corresponds to the case in which the two sets y_l and gx_k have non-empty overlap for *any* permutation g, i.e. the case in which $y_l \cap gx_k = \emptyset$ with probability zero.

We conclude this discussion of subsets and simplices with an application of the results of Section 3.1 to a question posed by Sperner. Let $A \in L(S)$, and suppose that all of the maximal elements of A have rank k; i.e. suppose that every face of A is contained in a k-face of A. Let $[A]_l$ denote the collection of all l-faces of A, for $0 \le l \le k$. Is there a lower bound for the number of l-faces $|[A]_l|$, given the number of k-faces (maximal faces) $|[A]_k|$? The L.Y.M. inequality gives one answer to this question.

Theorem 3.2.7 *Suppose* $A \in L(S)$ *such that every maximal element of* A *has rank* k. *For* $0 \le l \le k$,

$$|[A]_l| \ge \frac{k!(n-k)!}{l!(n-l)!}|[A]_k|.$$

Proof Let $B_l = P_l(S) - [A]_l$. If $y \in B_l$ then y cannot be contained in any $x \in A$. In other words, the set $[A]_k \cup B_l$ is an antichain. It then follows from (3.1) that

$$\frac{|[A]_k|}{\binom{n}{k}} + \frac{|B_l|}{\binom{n}{l}} \le 1.$$

Meanwhile,

$$|B_l| = |P_l(S) - [A]_l| = \binom{n}{l} - |[A]_l|,$$

so that

$$\frac{|[A]_k|}{\binom{n}{k}} + 1 - \frac{|[A]_l|}{\binom{n}{l}} \le 1.$$

It follows that

$$|[A]_l| \geq \binom{n}{l}\binom{n}{k}^{-1}|[A]_k| = \frac{k!(n-k)!}{l!(n-l)!}|[A]_k|.$$

□

For example, in the case $l = k - 1$, we have

$$|[A]_{k-1}| \geq \frac{k}{n-k+1}|[A]_k|.$$

3.3 A discrete analogue of Helly's theorem

We turn next to a special case and discrete analogue of Helly's theorem, whose classical geometric version is given in Chapter 5.

Theorem 3.3.1 (The discrete Helly theorem) *Let S be a finite set of size $|S| = n$, and let F be a family of subsets of S. Suppose that, for any subset $G \subseteq F$ such that $|G| \leq n$ (that is, every subfamily of cardinality at most n of F),*

$$\bigcap_{A \in G} A \neq \emptyset.$$

Then

$$\bigcap_{A \in F} A \neq \emptyset.$$

In other words, if every n elements of F have non-empty intersection, then the entire family F of subsets has non-empty intersection.

Proof If $|F| \leq n$ the result is trivial.

Suppose that Theorem 3.3.1 holds for $|F| = m$, for some $m \geq n$. We show that the theorem also holds for $|F| = m + 1$.

Write $S = \{s_1, \ldots, s_n\}$ and $F = \{A_1, \ldots, A_{m+1}\}$, where $A_i \subseteq S$. For each $1 \leq j \leq m + 1$ denote by L_j the intersection

$$L_j = \bigcap_{i \neq j} A_i.$$

Our induction assumption for the case $|F| = m$ implies that each L_j is non-empty. Therefore, there exists $s_{i_j} \in L_j$ for each j. However, since $m + 1 > n$ there must be $s \in S$ such that $s \in L_{j_1} \cap L_{j_2}$ for some $j_1 \neq j_2$. Since $s \in L_{j_1}$, we have $s \in A_i$ for all $i \neq j_1$. Similarly, $s \in A_i$ for all

$i \neq j_2$. Hence,

$$s \in \bigcap_{i=1}^{m+1} A_i = \bigcap_{A \in F} A.$$

$\square$

3.4 Notes

For a general reference to the theory of lattices and partially ordered sets, see [92] (also [3, 27, 42, 80]). For a graph-theoretic viewpoint, see [7]. In [84], Schanuel developed the Euler characteristic from a category-theoretic perspective. In [12], Chen extended the Euler characteristic to linear combinations of indicator functions of unbounded closed convex sets and unbounded relatively open convex sets. A thorough treatment of the combinatorial theory of the Euler characteristic appeared in [79, 80].

The face enumerators μ_i play a role analogous to that of the *intrinsic volumes* on parallelotopes and polyconvex sets (see Sections 4.2 and 7.2). The discrete basis theorem and kinematic formulas can be extended to the more general setting of finite vector spaces; see [53].

Sperner's Theorem 3.1.1 first appeared in [91], as did Theorem 3.2.7. The proof presented in this section is due to Lubell [63]. Meshalkin proved Theorem 3.1.5 in [73]. In this section we follow a more simplified approach due to Hochberg and Hirsch [45]. The generalization of Sperner's theorem to r-families (Theorem 3.1.2) was originally due to Erdős [22], but the proof given in this section is due to Harper and Rota [42].

While Theorem 3.2.7 gives a lower bound to the number of l-faces of a simplicial complex A, all of whose maximal elements have dimension k, Katona [48] and Kruskal [57] independently obtained a much stronger result, giving the exact minimum size of $[A]_l$, a minimum that is independent of the size of the ambient set S. Katona's original paper may also be found in [27]. For a shorter proof of the Katona–Kruskal theorem, see [3, 7]. A survey of results in extremal set theory, including generalizations of Sperner's theorem and the L.Y.M. inequality, was presented in [29].

The discrete Helly Theorem 3.3.1 is a special case of a more general theorem of convex geometry. This geometric result is treated in greater detail in Chapter 5 (see also [85, pp. 3–5]). Helly's original theorem motivated considerable developments in the field of combinatorial geometry, and 'Helly-type' theorems are now common in many branches of mathematics [7, 18, 21].

4

The intrinsic volumes for parallelotopes

We develop next a theory of invariant valuations for the lattice of finite unions of orthogonal parallelotopes, having edges parallel to a fixed frame. Many of the central results in geometric probability can be stated and proven easily in the context of parallelotopes. For this reason the lattice of parallelotopes will serve as a model for the more difficult task of developing a lattice theory for finite unions of compact convex sets in $\mathbf{R}^n$. The Euler characteristic, intrinsic volumes, and valuation characterization theorems for the lattice of parallelotopes will serve as prototypes in Chapters 5–9 for analogous constructions and characterization theorems in the more general context of polyconvex sets.

4.1 The lattice of parallelotopes

Choose a Cartesian coordinate system in $\mathbf{R}^n$, which shall remain fixed throughout this chapter, and let Par(n) denote the family of sets that are obtained by taking finite unions and intersections of orthogonal parallelotopes (i.e. rectilinear boxes), with sides parallel to the coordinate axes. If $P \in$ Par(n), we shall say that P is of dimension n (or has *full dimension*) if P is not contained in a finite union of hyperplanes of $\mathbf{R}^n$; that is, if P has a non-empty interior. Otherwise, we shall say that P is of *lower dimension*. (Recall that a *hyperplane* in $\mathbf{R}^n$ is a plane of dimension $n - 1$, not necessarily through the origin.) In general, a set $P \in$ Par(n) has dimension k if P is contained in a finite union of k-planes in $\mathbf{R}^n$, but is not contained in any finite union of $k - 1$ planes.

Note that Par(n) is closed under finite unions and intersections. This follows from the basic fact that the intersection of two parallelotopes is a parallelotope. In other words, Par(n) is a distributive lattice.

Denote by $\widetilde{T}_n$ the group generated by translations and permutations of coordinates in $\mathbf{R}^n$. For $A \subseteq \mathbf{R}^n$ and $g \in \widetilde{T}_n$, write

$$gA = g(A) = \{g(a) : a \in A\}.$$

A valuation μ defined on $\mathrm{Par}(n)$ is said to be *invariant* when

$$\mu(gP) = \mu(P) \tag{4.1}$$

for all $g \in \widetilde{T}_n$ and all $P \in \mathrm{Par}(n)$. If $\mu(gP) = \mu(P)$ is known to hold only for translations g of $\mathbf{R}^n$, then we shall say that μ is *translation invariant*.

The object of this section is to determine all invariant valuations defined on $\mathrm{Par}(n)$. To avoid pathological cases, we shall impose a continuity condition on the valuations to be considered.

For $A \subseteq \mathbf{R}^n$ and $x \in \mathbf{R}^n$, the distance $d(x, A)$ from the point x to the set A is given by

$$d(x, A) = \inf_{a \in A} d(x, a),$$

where $d(x, a) = |x - a|$ is the usual distance between points in $\mathbf{R}^n$. Note that $d(x, A) = 0$ if $x \in A$ or if x is a limit point of A.

For $K, L \subseteq \mathbf{R}^n$, the *Hausdorff distance* $\delta(K, L)$ is defined by

$$\delta(K, L) = \max \left(\sup_{a \in K} d(a, L), \sup_{b \in L} d(b, K) \right). \tag{4.2}$$

If K and L are compact, then $\delta(K, L) = 0$ if and only if $K = L$. A sequence of compact subsets K_n of $\mathbf{R}^n$ *converges* to a set K, or $K_n \longrightarrow K$, if $\delta(K_n, K) \longrightarrow 0$ as $n \to \infty$.

Let B_n denote the unit ball in $\mathbf{R}^n$. For $K \subseteq \mathbf{R}^n$ compact and $\epsilon > 0$, define

$$K + \epsilon B_n = \{x + \epsilon u : x \in K \text{ and } u \in B_n\}.$$

The following lemma gives an easier and more practical way to think about the distance δ.

Lemma 4.1.1 *Let $K, L \subseteq \mathbf{R}^n$ be compact sets. Then $\delta(K, L) \le \epsilon$ if and only if $K \subseteq L + \epsilon B_n$ and $L \subseteq K + \epsilon B_n$.*

Proof Suppose that $K \subseteq L + \epsilon B_n$. For $x \in K$ there exist $y \in L$ and $u \in B_n$ such that $x = y + \epsilon u$. In other words, $|x - y| \le \epsilon$, so that $d(x, L) \le \epsilon$. Similarly, if $L \subseteq K + \epsilon B_n$, then $d(y, K) \le \epsilon$ for all $y \in L$. Therefore, $\delta(K, L) \le \epsilon$.

Meanwhile, if there exists $x \in K$ such that $x \notin L + \epsilon B_n$ (or vice versa), then $|x - y| > \epsilon$ for all $y \in L$, so that $\delta(K, L) \geq d(x, L) > \epsilon$. $\square$

In view of Lemma 4.1.1, we see that a sequence of compact subsets $K_n \longrightarrow K$, if for $\epsilon > 0$ there exists $N > 0$ such that $K \subseteq K_i + \epsilon B_n$ and $K_i \subseteq K + \epsilon B_n$ whenever $i > N$.

Theorem 4.1.2 *The distance δ defines a metric on the set of all compact subsets of $\mathbf{R}^n$.*

Proof The distance δ is clearly symmetric and non-negative. To verify the triangle inequality, suppose that $K, L, M \subseteq \mathbf{R}^n$ are compact. Let $\epsilon_1 = \delta(K, M)$ and $\epsilon_2 = \delta(L, M)$. By Lemma 4.1.1, $K \subseteq M + \epsilon_1 B_n$ and $M \subseteq L + \epsilon_2 B_n$, so that

$$K \subseteq L + \epsilon_2 B_n + \epsilon_1 B_n = L + (\epsilon_2 + \epsilon_1) B_n.$$

Similarly, $L \subseteq K + (\epsilon_2 + \epsilon_1) B_n$. Lemma 4.1.1 then implies that $\delta(K, L) \leq \epsilon_1 + \epsilon_2$. $\square$

We now focus once again on the lattice Par(n) of finite unions of parallelotopes. A valuation μ is said to be *continuous* on Par(n), provided that

$$\mu(P_i) \longrightarrow \mu(P)$$

whenever P_i, P are *parallelotopes* (and *not* just finite unions) and $P_i \longrightarrow P$.

Another condition that will prove useful is monotonicity. A valuation μ is said to be *increasing* on Par(n), provided that $\mu(P) \leq \mu(Q)$, whenever $P, Q \in$ Par(n) and $P \subseteq Q$. Similarly one defines *decreasing* valuations. A valuation μ is said to be *monotone* on Par(n), if μ is either an increasing valuation or a decreasing valuation.

When studying valuations on Par(n) we may restrict our attention to the generating set of parallelotopes in $\mathbf{R}^n$ with edges parallel to the coordinate axes. Specifically, we have the following extension theorem.

Theorem 4.1.3 (Groemer's extension theorem for Par(n)) *A valuation μ defined on parallelotopes with edges parallel to the coordinate axes admits a unique extension to a valuation on the lattice Par(n).*

Proof In view of Groemer's integral Theorem 2.2.1, it is sufficient to show that μ defines an integral on the space of indicator functions of parallelotopes.

The proposition is trivial in dimension zero. Assume that the proposition holds in dimension $n - 1$. Suppose that there exist distinct parallelotopes $P_1, \ldots, P_m$ such that

$$\sum_{i=1}^{m} \alpha_i I_{P_i} = 0 \tag{4.3}$$

while

$$\sum_{i=1}^{m} \alpha_i \mu(P_i) = 1. \tag{4.4}$$

Let k be the number of parallelotopes P_i in the expressions above of full dimension n, and suppose that k is minimal over all possible such expressions.

If $k = 0$ then $P_1, \ldots, P_m$ are each contained inside a hyperplane. Let l denote the (finite) number of hyperplanes containing the parallelotopes $P_1, \ldots, P_m$, and suppose that l is minimal over all such expressions.

If $l = 1$ then $P_1, \ldots, P_m$ are all contained in a single hyperplane. It then follows from the induction assumption (on the dimension of the ambient Euclidean space) that we have a contradiction. Therefore $l > 1$, and there exist hyperplanes $H_1, \ldots, H_l$, orthogonal to the coordinate axes, such that $P_i \subseteq H_1 \cup \cdots \cup H_l$ for $i = 1, \ldots, m$. Suppose, without loss of generality, that $P_1 \subseteq H_1$.

Since $I_{P_i \cap H_1} = I_{P_i} I_{H_1}$, it follows from (4.3) that

$$\sum_{i=1}^{m} \alpha_i I_{P_i \cap H_1} = 0. \tag{4.5}$$

Meanwhile,

$$\sum_{i=1}^{m} \alpha_i \mu(P_i \cap H_1) = 0, \tag{4.6}$$

by the induction assumption on dimension, since each $P_i \cap H_1 \subseteq H_1$, a hyperplane.

Subtracting equations (4.5) and (4.6) from (4.3) and (4.4) respectively, we have

$$\sum_{i=1}^{m} \alpha_i (I_{P_i} - I_{P_i \cap H_1}) = 0, \tag{4.7}$$

and

$$\sum_{i=1}^{m} \alpha_i (\mu(P_i) - \mu(P_i \cap H_1)) = 1. \tag{4.8}$$

Since $P_1 \cap H_1 = P_1$, equations (4.7) and (4.8) take the form of (4.3) and (4.4), where the nonzero terms involve parallelotopes P_i in at most $l - 1$ hyperplanes, contradicting the minimality of l.

It follows that $k \geq 1$. Suppose then that P_1 has dimension n. Choose a hyperplane H, with associated closed half-spaces H^+ and H^- such that $P_1 \cap H$ is a facet of P_1, oriented so that $P_1 \subset H^+$. Since $I_{P_i \cap H^+} = I_{P_i} I_{H^+}$, it follows from (4.3) that

$$\sum_{i=1}^{m} \alpha_i I_{P_i \cap H^+} = 0.$$

Similarly,

$$\sum_{i=1}^{m} \alpha_i I_{P_i \cap H} = 0 \quad \text{and} \quad \sum_{i=1}^{m} \alpha_i I_{P_i \cap H^-} = 0.$$

Meanwhile, since μ is a valuation,

$$\sum_{i=1}^{m} \alpha_i \mu(P_i) = \sum_{i=1}^{m} \alpha_i \mu(P_i \cap H^+) + \sum_{i=1}^{m} \alpha_i \mu(P_i \cap H^-) - \sum_{i=1}^{m} \alpha_i \mu(P_i \cap H).$$

Since the sets $P_i \cap H$ lie inside a space of dimension $n - 1$, the sum $\sum_{i=1}^{m} \alpha_i \mu(P_i \cap H) = 0$ by the induction assumption. Because $P_1 \cap H^-$ has dimension $n - 1$, the sum $\sum_{i=1}^{m} \alpha_i \mu(P_i \cap H^-) = 0$ by the minimality of k. From (4.4) we have

$$\sum_{i=1}^{m} \alpha_i \mu(P_i \cap H^+) = \sum_{i=1}^{m} \alpha_i \mu(P_i) = 1.$$

There are n hyperplanes $H_1, \ldots, H_n$ such that

$$P_1 = \bigcap_{i=1}^{n} H_i^+.$$

By iterating the preceding argument, we have

$$\sum_{i=1}^{m} \alpha_i \mu(P_i \cap H_1^+ \cap \cdots \cap H_n^+) = 1,$$

so that

$$\sum_{i=1}^{m} \alpha_i \mu(P_i \cap P_1) = 1,$$

while a similar argument using (4.3) gives

$$\sum_{i=1}^{m} \alpha_i I_{P_i \cap P_1} = 0.$$

After repeating this argument with parallelotopes $P_2, \ldots, P_m$ we have

$$\sum_{i=1}^{m} \alpha_i \mu(P_1 \cap \cdots \cap P_m) = \left(\sum_{i=1}^{m} \alpha_i\right) \mu(P_1 \cap \cdots \cap P_m) = 1.$$

This implies that $\alpha_1 + \cdots + \alpha_m \neq 0$ and that $P_1 \cap \cdots \cap P_m \neq \emptyset$. Meanwhile a similar argument using (4.3) gives

$$\sum_{i=1}^{m} \alpha_i I_{P_1 \cap \cdots \cap P_m} = \left(\sum_{i=1}^{m} \alpha_i\right) I_{P_1 \cap \cdots \cap P_m} = 0,$$

so that either $\alpha_1 + \cdots + \alpha_m = 0$ or $P_1 \cap \cdots \cap P_m = \emptyset$, a contradiction in either case.   □

4.2 Invariant valuations on parallelotopes

To begin the classification of invariant valuations on Par(n), consider the problem in $\mathbf{R}^1$. An element of Par(1) is a finite union of closed intervals. Set

$$\mu_0^1(A) = \text{number of connected components of } A,$$

$$\mu_1^1(A) = \text{length of } A.$$

One easily verifies that μ_0^1 and μ_1^1 are both continuous invariant valuations on Par(1). We shall prove that every continuous invariant valuation on Par(1) is a linear combination of μ_0^1 and μ_1^1.

Suppose that μ is a continuous invariant valuation on Par(1). Let $c = \mu(A)$, where A is a set consisting of a single point in $\mathbf{R}^1$, and let $\mu' = \mu - c\mu_0^1$. Note that the invariant valuation μ' vanishes on points. Define a continuous function $f : [0, +\infty) \longrightarrow [0, +\infty)$ by the equation

$$f(x) = \mu'([0, x]).$$

If A is a closed interval of length x, then the invariance of μ' implies that $\mu'(A) = f(x)$. If A and B are closed intervals of length x and y, such that $A \cap B$ is a point, then

$$f(x + y) = \mu'(A \cup B) = \mu'(A) + \mu'(B) - \mu'(A \cap B) = f(x) + f(y),$$

so that $f(x) = rx$ for some constant r. Hence, $\mu' = r\mu_1^1$, and our assertion is proved.

We now turn to $\mathbf{R}^n$. There is one well known continuous invariant valuation defined on Par(n), namely, the volume. Denote by $\mu_n(P)$ the volume of a finite union P of parallelotopes of dimension n.

Recall that the *elementary symmetric functions* of $x_1, x_2, \ldots, x_n$ are the polynomials

$$e_0 = 1,$$

$$e_k(x_1, x_2, \ldots, x_n) = \sum_{1 \leq i_1 < \cdots < i_k \leq n}^{n} x_{i_1} x_{i_2} \cdots x_{i_k}, \ 1 \leq k \leq n.$$

We shall prove the following theorem.

Theorem 4.2.1 *For $0 \leq k \leq n$, there exists a unique continuous valuation μ_k on $\mathrm{Par}(n)$, invariant under translations and permutations of coordinates, such that*

$$\mu_k(P) = e_k(x_1, x_2, \ldots, x_n) \tag{4.9}$$

whenever P is a parallelotope with sides of length $x_1, x_2, \ldots, x_n$.

For example, $\mu_n(P)$ is the volume of P, if P has dimension n.

Proof Let μ_0^1 and μ_1^1 be the valuations previously defined on $\mathbf{R}^1$, and let $\mu_t^1 = \mu_0^1 + t\mu_1^1$, where t is a variable. Consider the product valuation on $\mathbf{R}^n$, given by the n-fold product

$$\mu_t^n = \mu_t^1 \times \mu_t^1 \times \cdots \times \mu_t^1. \tag{4.10}$$

Because a parallelotope in $\mathrm{Par}(n)$ is a Cartesian product of line segments, μ_t^n defines a valuation on *parallelotopes*. By Groemer's extension Theorem 4.1.3, μ_t^n then extends to a valuation on all of $\mathrm{Par}(n)$.

Let us compute the value of μ_t^n on the parallelotope P with edges of length $x_1, x_2, \ldots, x_n$. Since $P = I_1 \times I_2 \times \cdots \times I_n$, where I_j is an interval of length x_j, the definition (4.10) implies that

$$\mu_t^n(P) = \mu_t^1(I_1)\mu_t^1(I_2) \cdots \mu_t^1(I_n). \tag{4.11}$$

Meanwhile

$$\mu_t^1(I_j) = \mu_0^1(I_j) + t\mu_1^1(I_j) = 1 + tx_j.$$

By expanding the right-hand side of (4.11), we obtain

$$\begin{aligned}
\mu_t^n(P) &= (1 + tx_1)(1 + tx_2) \cdots (1 + tx_n) \\
&= 1 + e_1(x_1, \ldots x_n)t + e_2(x_1, \ldots x_n)t^2 + \cdots + e_n(x_1, \ldots x_n)t^n.
\end{aligned}$$

Now let $Q \in \mathrm{Par}(n)$. The set Q can be expressed as a union $Q = P_1 \cup P_2 \cup \cdots \cup P_n$, where the P_i are parallelotopes. (Recall that the intersection of a collection of parallelotopes is a parallelotope.) By the inclusion–exclusion principle,

$$\mu_t^n(Q) = \sum_i \mu_t^n(P_i) - \sum_{i<j} \mu_t^n(P_i \cap P_j) + \cdots - \cdots.$$

Collecting terms in each power of t, we conclude that there exist valuations $\mu_0, \mu_1, \ldots, \mu_n$ such that

$$\mu_t^n(Q) = \mu_0(Q) + t\mu_1(Q) + t^2\mu_2(Q) + \cdots + t^n\mu_n(Q),$$

where the valuations μ_i are uniquely determined by the identities (4.9).

$\square$

Note that, if the parallelotope P has dimension $k < n$, then the valuation $\mu_i(P)$ has an ambiguous sense. It may indicate the value of $\mu_i(P)$ in $\mathbf{R}^n$ as defined in the previous theorem, but it may also denote the value of $\mu_i(P)$ when computed within a lower dimensional plane containing P (and isomorphic to $\mathbf{R}^m$, where $k \le m < n$). It is a notable consequence of the preceding theorem that the two valuations coincide. In other words,

$$\mu_i^m(P) = \mu_i^n(P).$$

Therefore, there will be no need to indicate the dependence of $\mu_i(P)$ on the space $\mathbf{R}^n$ in which the parallelotope P is embedded. We summarize this remarkable fact as follows.

Theorem 4.2.2 *The valuations μ_i on* Par(n) *are normalized independently of the dimension n.* $\square$

In other words, the value $\mu_k(P)$ is 'intrinsic' to the set P, and independent of the dimension of the ambient space. For this reason the valuation μ_k is called the *kth intrinsic volume*.

The valuation μ_0 is called the *Euler characteristic*. As a corollary of Theorem 4.2.1, we see that the Euler characteristic is the only valuation on Par(n) that takes the value 1 on all (non-empty) parallelotopes.

Let H_1 and H_2 be complementary orthogonal subspaces of $\mathbf{R}^n$ spanned by subsets of the given coordinate system and having dimensions h and $n - h$, respectively. Let P_i be a parallelotope in H_i and let $P = P_1 \times P_2$. The intrinsic volumes satisfy the following property with regard to orthogonal Cartesian products.

Proposition 4.2.3

$$\mu_i(P_1 \times P_2) = \sum_{r+s=i} \mu_r(P_1)\mu_s(P_2). \qquad (4.12)$$

The identity (4.12) is therefore valid when P_1 and P_2 are finite unions of parallelotopes.

Proof Suppose that P_1 has sides of length $x_1, \ldots, x_h$ and P_2 has sides of length $y_1, \ldots, y_{n-h}$. Then we have

$$\sum_{r+s=i} \mu_r(P_1)\mu_s(P_2)$$

$$= \sum_{r+s=i} \left(\sum_{1 \le j_1 < \cdots < j_r \le h} x_{j_1} \cdots x_{j_r} \sum_{1 \le k_1 < \cdots < k_s \le n-h} y_{k_1} \cdots y_{k_s} \right).$$

Let $j_{r+1} = k_1 + h, \ldots, j_i = j_{r+s} = k_s + h$, and let $x_{h+1} = y_1, \ldots, x_n = y_{n-h}$. Then

$$\sum_{r+s=i} \mu_r(P_1)\mu_s(P_2)$$

$$= \sum_{r+s=i} \left(\sum_{1 \le j_1 < \cdots < j_r \le h} x_{j_1} \cdots x_{j_r} \sum_{h+1 \le j_{r+1} < \cdots < j_i \le n} x_{j_{r+1}} \cdots x_{j_i} \right)$$

$$= \sum_{1 \le j_1 < \cdots < j_r < j_{r+1} < \cdots < j_i \le n} x_{j_1} \cdots x_{j_r} x_{j_{r+1}} \cdots x_{j_i}$$

$$= \mu_i(P_1 \times P_2).$$

$\square$

A valuation μ on $\mathrm{Par}(n)$ is said to be *simple* if $\mu(P) = 0$ for all P of dimension less than n. The restriction of the volume μ_n to the lattice $\mathrm{Par}(n)$ is characterized by the following theorem.

Theorem 4.2.4 (The volume theorem for $\mathrm{Par}(n)$) *Let μ be a translation invariant simple valuation defined on $\mathrm{Par}(n)$, and suppose that μ is either continuous or monotone. Then there exists $c \in \mathbf{R}$ such that $\mu(P) = c\mu_n(P)$ for all $P \in \mathrm{Par}(n)$; that is, μ is equal to the volume, up to a constant factor.*

Proof Let $[0,1]^n$ denote the unit cube in $\mathbf{R}^n$, and let $c = \mu([0,1]^n)$. Recall that μ is translation invariant and vanishes on lower dimensions. Since $\mu([0,1]^n) = c$, a simple cut-and-paste argument shows that

$\mu([0, 1/k]^n) = c/k^n$ for all integers $k > 0$. Therefore, $\mu(C) = c\mu_n(C)$ for every box C of rational dimensions, with sides parallel to the coordinate axes. This follows from the fact that such a box can be built up by stacking cubes of the form $[0, 1/k]^n$ for some $k > 0$. Since μ is either continuous or monotone, it follows that $\mu(C) = c\mu_n(C)$ for every box C of positive real dimensions, with sides parallel to the coordinate axes. It then follows from the inclusion–exclusion principle that $\mu(P) = c\mu_n(P)$ for all $P \in \text{Par}(n)$. □

The condition of either continuity or monotonicity is necessary to the characterization given by Theorem 4.2.4. If we omit these conditions then counterexamples to Theorem 4.2.4 can be found even in the case of Par(1)! To see this, recall that $\mathbf{R}$ is a vector space of infinite dimension over the field $\mathbf{Q}$ of rational numbers. Denote this vector space $\mathbf{R_Q}$. Since the dual space $\mathbf{R_Q^*}$ is also of infinite dimension, there exists a nontrivial map $f \in \mathbf{R_Q^*}$; i.e., a linear map $f : \mathbf{R_Q} \longrightarrow \mathbf{Q}$ such that $f(1) = 1$ and $f(x) \in \mathbf{Q}$ for all $x \in \mathbf{R}$.

A parallelotope $P \in \text{Par}(1)$ is just a closed bounded interval of the form $[a, b]$, having length $b - a$. Define a valuation η on parallelotopes (intervals) in Par(1) by the formula

$$\eta([a, b]) = f(b - a).$$

Evidently η is invariant, depending only on the length of the closed interval. Moreover, if $a \le c \le b \le d$, then

$$\begin{aligned}
\eta([a, b] &\cup [c, d]) + \eta([a, b] \cap [c, d]) \\
&= \eta([a, d]) + \eta([c, b]) \\
&= f(d - a) + f(b - c) \\
&= (f(c - a) + f(b - c) + f(d - b)) + f(b - c) \\
&= f(b - a) + f(d - c) \\
&= \eta([a, b]) + \eta([c, d]),
\end{aligned}$$

by the linearity of f. It now follows from Groemer's extension Theorem 4.1.3 that η extends to an invariant valuation on Par(1) that vanishes on lower dimensions. However, η is *not* equal to length (one-dimensional volume), since η takes only rational values. In other words, invariance alone is insufficient to characterize the volume – either continuity or monotonicity is also required. The reasoning that underlies this counterexample to Theorem 4.2.4 is easily extended to provide counterexamples for Par(n), where $n \ge 1$.

We are now able to determine all continuous valuations on $\mathrm{Par}(n)$ that are invariant under translation and permutations of coordinates. We shall not yet prove that they are also rotation invariant.

Theorem 4.2.5 *The valuations* $\mu_0, \mu_1, \ldots, \mu_n$ *form a basis for the vector space of all continuous invariant valuations defined on* $\mathrm{Par}(n)$.

Proof Let μ be a continuous invariant valuation on $\mathrm{Par}(n)$. Denote by $x_1, x_2, \ldots, x_n$ the standard orthonormal basis for $\mathbf{R}^n$, and let H_j denote the $(n-1)$-hyperplane in $\mathbf{R}^n$ spanned by the coordinate vectors $x_1, \ldots, x_{j-1}, x_{j+1}, \ldots, x_n$. The restriction of μ to H_j is an invariant valuation on parallelotopes in H_j. Proceeding by induction, we may assume that

$$\mu(A) = \sum_{i=0}^{n-1} c_i \mu_i(A),$$

for all $A \in \mathrm{Par}(n)$ such that $A \subseteq H_j$. Moreover, the coefficients c_i are the same for each choice of H_j, since the valuations $\mu, \mu_0, \ldots, \mu_{n-1}$ are invariant under permutation of the coordinates $x_1, \ldots, x_n$. Thus, the valuation

$$\mu - \sum_{i=0}^{n-1} c_i \mu_i$$

vanishes on all lower dimensional parallelotopes in $\mathrm{Par}(n)$, since any such parallelotope is contained in a hyperplane parallel to one of the hyperplanes H_j. By Theorem 4.2.4,

$$\mu - \sum_{i=0}^{n-1} c_i \mu_i = c_n \mu_n,$$

where μ_n is the volume on $\mathbf{R}^n$, and where c_n is a real constant. In other words,

$$\mu = \sum_{i=0}^{n} c_i \mu_i.$$

$\square$

A valuation μ on $\mathrm{Par}(n)$ is said to be *homogeneous* of degree $k > 0$ if

$$\mu(\alpha P) = \alpha^k \mu(P)$$

for all $P \in \mathrm{Par}(n)$ and all $\alpha \geq 0$.

Corollary 4.2.6 *Let μ be a continuous invariant valuation defined on* Par(n) *that is homogeneous of degree k, for some $0 \leq k \leq n$. Then there exists $c \in \mathbf{R}$ such that $\mu(P) = c\mu_k(P)$ for all $P \in$ Par(n).*

Proof By Theorem 4.2.5 there exist $c_1, \ldots, c_n \in \mathbf{R}$ such that

$$\mu = \sum_{i=0}^{n} c_i \mu_i.$$

If $P = [0,1]^n$ then, for $\alpha > 0$,

$$\mu(\alpha P) = \sum_{i=0}^{n} c_i \mu_i(\alpha P) = \sum_{i=0}^{n} c_i \alpha^i \mu_i(P) = \sum_{i=0}^{n} \binom{n}{i} c_i \alpha^i$$

Meanwhile,

$$\mu(\alpha P) = \alpha^k \mu(P) = \alpha^k \sum_{i=0}^{n} c_i \mu_i(P) = \sum_{i=0}^{n} \binom{n}{i} c_i \alpha^k$$

Therefore, $c_i = 0$ if $i \neq k$, and $\mu = c_k \mu_k$. □

4.3 Notes

For a more complete discussion of the Hausdorff topology on the space of compact subsets of $\mathbf{R}^n$ and the subspace of compact convex sets, see [85, pp. 47–61]. Theorem 4.1.3 is a special case of a more general extension theorem of Groemer, in which the lattice Par(n) is replaced with the lattice of polytopes in $\mathbf{R}^n$ (see [32]). Theorem 4.2.5 can be generalized to a classification of all continuous translation invariant valuations on the lattice Par(n), omitting the requirement that valuations be invariant under permutation of coordinates. The result is a 2^n-dimensional space of valuations, with a basis indexed by the collection of all *coordinate* subspaces of $\mathbf{R}^n$ with respect to the fixed basis for parallelotope edges in Par(n); that is, by the set of all 2^n subsets of that n-element basis. For a detailed treatment, see [54].

5

The lattice of polyconvex sets

We turn now to the lattice of polyconvex sets, which is a natural setting for the study of classical geometric probability. In Section 5.2 we define the Euler characteristic on polyconvex sets, which is an important tool for the extension in Chapter 7 of the intrinsic volumes of Section 4.2 to polyconvex sets. The Euler characteristic will also reappear in Chapter 10, in which we generalize the discrete kinematic formula of Chapter 3 to polyconvex sets. Section 5.5, while interesting in its own right, points to the correct normalization of the rotation invariant measures on Grassmannians in Section 6.1.

5.1 Polyconvex sets

A subset K of $\mathbf{R}^n$ is said to be *convex* if any two points x and y in K are the endpoints of a line segment lying inside K. Denote by $\mathcal{K}^n$ the collection of all compact convex subsets of $\mathbf{R}^n$. A finite union of compact convex sets will be called a *polyconvex* set (a term suggested by E. De Giorgi).

If A is a polyconvex set in $\mathbf{R}^n$, we shall say that A is of dimension n if A is not contained in a finite union of hyperplanes of $\mathbf{R}^n$; that is, if A has a non-empty interior. Otherwise, we shall say that A is of *lower dimension*. The union and intersection of polyconvex sets are polyconvex. This follows from the basic fact that the intersection of two convex sets is convex. In other words, the family of polyconvex sets in $\mathbf{R}^n$ is a distributive lattice. We denote this lattice Polycon(n). Note that Par(n) is a *sublattice* of Polycon(n). The lattice Polycon(n) is also sometimes called the *convex ring*.

A non-empty compact convex set $K \in \mathcal{K}^n$ is determined uniquely by its *support function* h_K : $\mathbf{S}^{n-1} \longrightarrow \mathbf{R}$, defined by $h_K(u) =$

$\max_{x \in K}\{x \cdot u\}$, where $\cdot$ denotes the standard inner product on $\mathbf{R}^n$. For example, if $v \in \mathbf{R}^n$ and $\bar{v}$ denotes the line segment with endpoints v and $-v$, then $h_{\bar{v}}(u) = |u \cdot v|$, for all $u \in \mathbf{S}^{n-1}$.

More generally, suppose that $h : \mathbf{S}^{n-1} \longrightarrow \mathbf{R}$, and consider the radial extension $\tilde{h} : \mathbf{R}^n \longrightarrow \mathbf{R}$ given by $\tilde{h}(\alpha u) = \alpha h(u)$, for all $u \in \mathbf{S}^{n-1}$ and $\alpha \geq 0$. The original function h is a support function of a compact convex set in $\mathbf{R}^n$ if and only if the radial extension $\tilde{h}$ is *sublinear*; that is,

$$h(x + y) \leq h(x) + h(y),$$

for all $x, y \in \mathbf{R}^n$.

Note that, for $u \in \mathbf{S}^{n-1}$, a compact convex set K lies entirely on one side (the '$-u$' side) of the hyperplane $H(K, u)$ determined by the equation $x \cdot u = h_K(u)$. The hyperplane $H(K, u)$ is called the *support plane* of K in the direction u. If $H(K, u)^-$ denotes the closed half-space $x \cdot u \leq h_K(u)$ bounded by $H(K, u)$, then we have

$$K = \bigcap_{u \in \mathbf{S}^n} H(K, u)^-.$$

For compact convex sets K and L the *Minkowski sum* $K + L$ is defined by

$$K + L = \{x + y : x \in K \text{ and } y \in L\}. \tag{5.1}$$

It is not difficult to show that $h_{K+L} = h_K + h_L$.

Recall from Lemma 4.1.1 that for compact sets K and L in $\mathbf{R}^n$, the Hausdorff metric satisfies $d(K, L) \leq \epsilon$ if and only if $K \subseteq L + \epsilon B$ and $L \subseteq K + \epsilon B$. It follows easily from (5.1) that

$$\delta(K, L) = \sup_{u \in \mathbf{S}^{n-1}} |h_K(u) - h_L(u)|. \tag{5.2}$$

In other words, the Hausdorff topology on $\mathcal{K}^n$ is also given by the uniform metric topology on the set of support functions of compact convex sets.

Denote by E_n the Euclidean group on $\mathbf{R}^n$; that is, the group generated by translations and (proper or improper) rotations. If $A \subseteq \mathbf{R}^n$ and $g \in E_n$, write

$$gA = g(A) = \{g(a) : a \in A\}.$$

The subgroup of translations (relative to a fixed Cartesian coordinate system) shall be denoted by T_n.

A valuation μ defined on polyconvex sets in $\mathbf{R}^n$ is said to be *rigid motion invariant* (or simply *invariant*, when no confusion is possible) if

$$\mu(A) = \mu(gA) \tag{5.3}$$

for all $g \in E_n$ and all $A \in \text{Polycon}(n)$. If the equality (5.3) holds only when $g \in T_n$, we say that μ is *translation invariant*.

Our objective is to determine all invariant valuations defined on polyconvex sets in $\mathbf{R}^n$. As with $\text{Par}(n)$, we impose a continuity condition on the valuations to be considered. A valuation μ is said to be *convex-continuous* (or simply *continuous*, when no confusion is possible) provided that

$$\mu(A_n) \longrightarrow \mu(A)$$

whenever A_n, A are compact *convex* sets and $A_n \longrightarrow A$ with respect to the metric (5.2). Examples of continuous invariant valuations on $\text{Polycon}(n)$ include *volume* and *surface area*. The following proposition shows that we can restrict our attention to convex-continuous valuations defined on the generating set $\mathcal{K}^n$.

Theorem 5.1.1 (Groemer's extension theorem for $\text{Polycon}(n)$)
A convex-continuous valuation μ on $\mathcal{K}^n$ admits a unique extension to a valuation on the lattice $\text{Polycon}(n)$.

Proof Suppose that μ is a continuous valuation. In view of Groemer's integral Theorem 2.2.1, it is sufficient to show that μ defines an integral on the space of indicator functions.

The theorem is trivial in dimension zero. Assume that the theorem holds in dimension $n-1$. Suppose that there exist distinct $K_1, \ldots, K_m \in \mathcal{K}^n$ such that

$$\sum_{i=1}^{m} \alpha_i I_{K_i} = 0 \tag{5.4}$$

while

$$\sum_{i=1}^{m} \alpha_i \mu(K_i) = 1. \tag{5.5}$$

Let m be the least positive integer for which such expressions (5.4) and (5.5) exist.

Choose a hyperplane H, with associated closed half-spaces H^+ and H^- such that $K_1 \subset \text{Int } H^+$. Since $I_{K_i \cap H^+} = I_{K_i} I_{H^+}$, it follows from (5.4) that

$$\sum_{i=1}^{m} \alpha_i I_{K_i \cap H^+} = 0.$$

Similarly,

$$\sum_{i=1}^{m} \alpha_i I_{K_i \cap H} = 0 \quad \text{and} \quad \sum_{i=1}^{m} \alpha_i I_{K_i \cap H^-} = 0.$$

Meanwhile, since μ is a valuation,

$$\sum_{i=1}^{m} \alpha_i \mu(K_i) = \sum_{i=1}^{m} \alpha_i \mu(K_i \cap H^+) + \sum_{i=1}^{m} \alpha_i \mu(K_i \cap H^-) - \sum_{i=1}^{m} \alpha_i \mu(K_i \cap H).$$

Since the sets $K_i \cap H$ lie inside a space of dimension $n - 1$, the sum $\sum_{i=1}^{m} \alpha_i \mu(K_i \cap H) = 0$ by the induction assumption. Because $K_1 \cap H^- = \emptyset$, the sum $\sum_{i=1}^{m} \alpha_i \mu(K_i \cap H^-) = 0$ by the minimality of m. From (5.5) we have

$$\sum_{i=1}^{m} \alpha_i \mu(K_i \cap H^+) = \sum_{i=1}^{m} \alpha_i \mu(K_i) = 1.$$

Choose a sequence of hyperplanes $H_1, H_2, \ldots$ such that $K_1 \subset \text{Int } H_i^+$ and

$$K_1 = \bigcap H_i^+.$$

By iterating the preceding argument, we have

$$\sum_{i=1}^{m} \alpha_i \mu(K_i \cap H_1^+ \cap \cdots \cap H_q^+) = 1$$

for all $q \geq 1$. Since μ is continuous, the limit as $q \to \infty$ gives

$$\sum_{i=1}^{m} \alpha_i \mu(K_i \cap K_1) = 1,$$

while a similar argument using (5.4) gives

$$\sum_{i=1}^{m} \alpha_i I_{K_i \cap K_1} = 0.$$

After repeating this argument with the bodies $K_2, \ldots, K_m$ we have

$$\sum_{i=1}^{m} \alpha_i \mu(K_1 \cap \cdots \cap K_m) = \left(\sum_{i=1}^{m} \alpha_i \right) \mu(K_1 \cap \cdots \cap K_m) = 1.$$

This implies that $\alpha_1 + \cdots + \alpha_m \neq 0$ and that $K_1 \cap \cdots \cap K_m \neq \emptyset$. Meanwhile a similar argument using (5.4) gives

$$\sum_{i=1}^{m} \alpha_i I_{K_1 \cap \cdots \cap K_m} = \left(\sum_{i=1}^{m} \alpha_i \right) I_{K_1 \cap \cdots \cap K_m} = 0,$$

so that either $\alpha_1 + \cdots + \alpha_m = 0$ or $K_1 \cap \cdots \cap K_m = \emptyset$, a contradiction in either case.   □

5.2 The Euler characteristic

Next, we shall extend the valuation μ_0 to the entire distributive lattice Polycon(n). We have seen that μ_0 is a well-defined functional on sets that are finite unions and intersections of parallelotopes, and that $\mu_0(P) = 1$ if P is a non-empty parallelotope. These results motivate the following theorem.

Theorem 5.2.1 (The existence of the Euler characteristic)
  There exists a unique convex-continuous invariant valuation μ_0^n defined on the family Polycon(n) *of all polyconvex sets in $\mathbf{R}^n$, such that $\mu_0^n(K) = 1$ whenever K is a non-empty compact convex set.*

The valuation μ_0^n is again called the Euler characteristic.
Proof We proceed by induction on the dimension n, the case $n = 1$ having been established previously. By Theorems 2.2.1 and 5.1.1, it will suffice to establish the existence of a linear functional L_n defined on $\mathcal{K}^n$-simple functions, such that $L_n(I_K) = 1$ whenever K is a non-empty compact convex set.

For $n = 1$, set

$$L_1(f) = \sum_{x \in \mathbf{R}} (f(x) - f(x + 0)),$$

where $f(x + 0) = \lim_{a \to 0^+} f(x + a)$. The sum on the right-hand side is finite, and, for $f = I_K$, where K is an interval $[a, b]$, we have

$$L_1(I_K) = f(b) - f(b + 0) = 1.$$

Thus, $L_1(I_K) = \mu_0^1(K)$, so that

$$L_1(f) = \int f \, d\mu_0^1.$$

For arbitrary n, choose an orthogonal coordinate system $x_1, x_2, \ldots, x_n$. Given the first coordinate x, let H_x be the hyperplane parallel to the coordinates $x_2, \ldots, x_n$ and passing through the point $(x, 0, \ldots, 0)$.

Let $f = f(x_1, x_2, \ldots, x_n)$ be a simple function. The function $f_x(x_2, \ldots, x_n) = f(x, x_2, \ldots, x_n)$ is a simple function in H_x, and we assume that $L_{n-1}(f_x)$ has been defined in H_x, since H_x is isomorphic to $\mathbf{R}^{n-1}$. Set $F(x) = L_{n-1}(f_x)$, and set

$$L_n(f) = L_1(F).$$

Note that the function F is simple, so that the right-hand side is well defined.

If $f = I_K$, where K is a compact convex set, then f_x is the indicator function of the slice of K by the hyperplane at $x_1 = x$, and F is the indicator function of the projection of K onto the x_1-coordinate axis. It follows that $L_1(F) = 1$. Since

$$L_n(f) = \int f \, \mathrm{d}\mu_0^n,$$

for some valuation μ_0^n, it follows that μ_0^n is the desired valuation. $\quad\square$

Note that the Euler characteristic μ_0^n is *normalized*. In other words, if K is a polyconvex set of dimension k in $\mathbf{R}^n$, and if V is a plane of dimension j containing K, then $\mu_0^j(K)$ computed within V is equal to $\mu_0^n(K)$ computed in $\mathbf{R}^n$. This follows from the fact that $\mu_0^n(K)$ can be computed via the inclusion–exclusion principle after K has been expressed as a finite union of compact convex sets, whereas $\mu_0(K) = 1$ for all non-empty compact convex sets K in spaces of any (finite) dimension. For this reason we write μ_0 in place of μ_0^n.

The argument in the preceding proof can also be used to compute the Euler characteristic of a polytope. By Corollary 2.2.2, the valuation μ_0 extends uniquely to a valuation defined on the relative Boolean algebra generated by Polycon(n), a valuation that is again denoted by μ_0.

Consider the (smaller) distributive sublattice of Polycon(n) generated by compact convex polytopes. Recall that a convex polytope is the intersection of a finite collection of closed half-spaces. A *polytope* is a finite union of convex polytopes. Given a polytope P, the boundary ∂P is also a polytope (which is *not* the case for an arbitrary compact convex set). Therefore, $\mu_0(\partial P)$ is defined.

Theorem 5.2.2 *If P is a compact convex polytope of dimension $n > 0$, then*

$$\mu_0(\partial P) = 1 - (-1)^n.$$

Proof Using again the notation of Theorem 5.2.1, we note that $H_x \cap \partial P = \partial(H_x \cap P)$ if x is not a boundary point of πP, the orthogonal *projection* of P onto the line $H_x^\perp$. Let $F(x) = \mu_0(\partial(H_x \cap P))$, where μ_0 is taken in the space H_x.

For the case $n = 1$ we have $\mu_0(\partial P) = 2 = 1 - (-1)$, since ∂P consists of two distinct points. For $n > 1$ it follows from the induction hypothesis that

$$\mu_0(H_x \cap \partial P) = \mu_0(\partial(H_x \cap P)) = 1 - (-1)^{n-1},$$

when $x \in \pi P$ is not a boundary point of πP. Meanwhile, if $x \in \partial(\pi P)$, we have

$$\mu_0(H_x \cap \partial P) = 1,$$

since $H_x \cap P$ is a face of P (though possibly a single point). Finally,

$$\mu_0(H_x \cap \partial P) = 0,$$

when $H_x \cap \partial P = \emptyset$.

We can now compute

$$L_1(F(x)) = \sum_x (F(x) - F(x + 0)),$$

a sum that vanishes except at the two points, call them a and b (with $a < b$), where H_x touches the boundary of P. The right-hand side then reduces to

$$F(a) - F(a + 0) + F(b) - F(b + 0).$$

We compute

$$F(b + 0) = 0,$$

$$F(b) = 1,$$

$$F(a) = 1,$$

$$F(a + 0) = 1 - (-1)^{n-1}.$$

On adding, we find that

$$L_1(F(x)) = 1 - 1 + (-1)^{n-1} + 1 = 1 + (-1)^{n-1} = 1 - (-1)^n,$$

as desired   □

If P is a compact convex polytope of dimension k in $\mathbf{R}^n$, let V be the k-dimensional plane containing P. We denote by relint(P) the interior of P relative to the topology of V; that is, the relative interior of P.

Theorem 5.2.3 *Let P be a compact convex polytope of dimension k in* $\mathbf{R}^n$. *Then*

$$\mu_0(\mathrm{relint}(P)) = (-1)^k.$$

Proof Since μ_0 is normalized independently of the ambient space, we compute within the k-dimensional plane in $\mathbf{R}^n$ containing P. From Theorem 5.2.2 we have

$$\mu_0(\mathrm{relint(P)}) = \mu_0(P) - \mu_0(\partial P) = (-1)^k.$$

$\square$

We can now generalize Euler's formula to arbitrary (nonconvex) polytopes. To this end, we define the notion of a *system of faces F* of a polytope P. This will be a family with the following properties:

- Every element of F is a convex polytope.
- $\bigcup_{Q \in F} \mathrm{relint}(Q) = P$.
- If $Q, Q' \in F$ and $Q \neq Q'$, then $\mathrm{relint}(Q) \cap \mathrm{relint}(Q') = \emptyset$

We can now prove the following result.

Theorem 5.2.4 (The Euler–Schläfli–Poincaré formula) *Let F be a system of faces of a polytope P, and let f_i be the number of elements of F of dimension i. Then*

$$\mu_0 = f_0 - f_1 + f_2 - \cdots$$

First proof Place a linear ordering on the elements of F, or 'faces', such that, if $Q < Q'$ then $\dim(Q) \leq \dim(Q')$. Evidently,

$$I_P = \sum_{Q \in F} (I_Q - I_{Q_1}),$$

where $Q_1 = Q \cap \left(\bigcup_{Q' < Q} Q' \right)$. Note that $I_Q - I_{Q_1} = I_{Q - Q_1}$ is the indicator of the set of points of $\mathbf{R}^n$ belonging to the face Q, but not to any of the preceding faces.

The sum on the right-hand side can be simplified as follows. Let F_i be the set of all faces of P of dimension i. Then

$$
\int_P \mathrm{d}\mu_0 = \sum_{Q \in F_0} \int (I_Q - I_{Q_1}) \, \mathrm{d}\mu_0 + \sum_{Q \in F_1} \int (I_Q - I_{Q_1}) \, \mathrm{d}\mu_0
$$
$$
+ \sum_{Q \in F_2} \int (I_Q - I_{Q_1}) \, \mathrm{d}\mu_0 + \cdots
$$

For $Q \in F_0$, we have $I_{Q_1} = 0$, so that

$$\sum_{Q \in F_0} \int (I_Q - I_{Q_1}) \, d\mu_0 = f_0.$$

For $Q \in F_i$, where $i > 0$, we have

$$Q_1 = Q \cap \left(\bigcup_{Q' < Q} Q' \right) = \partial Q,$$

and hence,

$$\begin{aligned}
\int (I_Q - I_{Q_1}) \, d\mu_0 &= \mu_0(Q) - \mu_0(\partial Q) = 1 - (1 - (-1)^i) \\
&= 1 - 1 + (-1)^i = (-1)^i.
\end{aligned}$$

Therefore,

$$\sum_{Q \in F_i} \int (I_Q - I_{Q_1}) \, d\mu_0 = (-1)^i f_i,$$

so that

$$\mu_0(P) = \int I_P \, d\mu_0 = f_0 - f_1 + f_2 - \cdots .$$

$\square$

Second proof We use the relative interior theorem. This requires extending the valuation μ_0 to the relative Boolean algebra generated by polyconvex sets. If F is any system of faces, then

$$\begin{aligned}
\mu_0(P) &= \mu_0 \left(\bigcup_{Q \in F} \mathrm{relint}(Q) \right) = \sum_{Q \in F} \mu_0(\mathrm{relint}(Q)) \\
&= \sum_{Q \in F} (-1)^{\dim Q}
\end{aligned}$$

by Theorem 5.2.3, since each $Q \in F$ is a convex polytope. We collect terms of the same dimension, and the theorem then follows. $\square$

5.3 Helly's theorem

The following is a remarkable application of the Euler characteristic.

Theorem 5.3.1 (Klee's theorem) *Let F be a finite family of compact convex sets such that*

$$\bigcup_{K \in F} K$$

is convex. Let $i < |F|$, and suppose that, for any subset $G \subseteq F$ such that $|G| = i$ (that is, every subset of cardinality i of F),

$$\bigcap_{K \in G} K \neq \emptyset.$$

Then there exists a subset H of F with cardinality $i + 1$, such that

$$\bigcap_{K \in H} K \neq \emptyset.$$

Proof Let $n > 1$ be a positive integer. Recall that

$$1 - \binom{n}{1} + \binom{n}{2} - \binom{n}{3} + \cdots + (-1)^j \binom{n}{j} \neq 0, \qquad (5.6)$$

for all positive integers $j < n$. To see why (5.6) holds, suppose that $j \leq \langle n/2 \rangle$. In this case the left-hand side of (5.6) is an alternating sum of *strictly increasing* terms, and is therefore unequal to zero. Since

$$\binom{n}{k} = \binom{n}{n-k}$$

and

$$1 - \binom{n}{1} + \binom{n}{2} - \binom{n}{3} + \cdots + (-1)^n \binom{n}{n} = (1-1)^n = 0$$

it also follows that (5.6) holds if we replace $j \leq \langle n/2 \rangle$ by $j \geq \langle n/2 \rangle$.

Now let $n = |F|$. From Theorem 5.2.1 and the inclusion–exclusion formula we have

$$1 = \mu_0 \left(\bigcup_{K \in F} K \right) = \sum_{K \in F} \mu_0(K) - \sum_{K \neq L \in F} \mu_0(K \cap L) + \cdots$$

$$= \binom{n}{1} - \binom{n}{2} + \binom{n}{3} - \cdots + (-1)^{i+1} \binom{n}{i}$$

whenever $\bigcap_{K \in G} K = \emptyset$ for all $G \subseteq F$ such that $|G| = i + 1$. However, this is impossible, by virtue of the inequality (5.6). $\square$

Given a set $A \subseteq \mathbf{R}^n$, the *convex hull* of A is the smallest convex set in $\mathbf{R}^n$ that contains A; that is, the intersection of all convex sets containing

A. The following is a simple and yet fundamental property of convex hulls in $\mathbf{R}^n$.

Theorem 5.3.2 (Carathéodory's theorem) *Let T be the convex hull of a family of compact convex sets $K_1, K_2, \ldots, K_m$ in $\mathbf{R}^n$. For each $x \in T$, there exists a subfamily $K_{j_1}, \ldots, K_{j_p}$, with convex hull $\overline{T}_x$, such that $p \leq n+1$ and $x \in \overline{T}_x$.*

Proof If $n = 0$ the result is trivial. Assuming that the theorem holds for dimension $n - 1$, we prove the theorem for dimension n.

Let $x \in T$. If x lies on the boundary of the compact convex set T, let H denote a support plane of T at x. Because H supports the convex set T, all of T lies inside one of the closed half-spaces bounded by H. Since $x \in T \cap H$, it follows that x lies in the convex hull of the convex sets $K_j \cap H$. By the induction assumption on dimension, there exists a subfamily $K_{j_1} \cap H, \ldots, K_{j_p} \cap H$ with convex hull T^* such that $p \leq n$ and $x \in T^*$. It follows that x lies in the convex hull of the subfamily $K_{j_1}, \ldots, K_{j_p}$.

If x lies in the interior of T, suppose that $x \notin K_j$ for any j (otherwise the proof is finished). Let ℓ denote a line through x that also meets K_m, and let x' be the point of intersection of ℓ with the boundary of T. Since ℓ meets the boundary of T at two points, choose x' so that x lies between x' and $\ell \cap K_m$. It follows from the argument in the previous paragraph that x' lies in the convex hull of a subfamily $K_{j_1}, \ldots, K_{j_p}$ of F, where $p \leq n$. It then follows that x must lie the convex hull of the sets $K_{j_1}, \ldots, K_{j_p}, K_m$, a collection of at most $n + 1$ sets in F. □

Combining Carathéodory's Theorem 5.3.2 with Klee's Theorem 5.3.1 we obtain the following celebrated theorem of Helly.

Theorem 5.3.3 (Helly's theorem) *Let F be a finite family of compact convex sets in $\mathbf{R}^n$. Suppose that, for any subset $G \subseteq F$ such that $|G| \leq n + 1$ (that is, every subset of cardinality at most $n + 1$ of F),*

$$\bigcap_{K \in G} K \neq \emptyset.$$

Then

$$\bigcap_{K \in F} K \neq \emptyset.$$

In other words, if every $n + 1$ elements of F have non-empty intersection, then the entire family F of convex sets has non-empty intersection.

Proof If $|F| \leq n + 1$ the result is trivial. Suppose that Theorem 5.3.3 holds for $|F| = m$, for some $m \geq n + 1$. We show that the theorem also holds for $|F| = m + 1$.

Let $F = \{K_1, \ldots, K_{m+1}\}$. For each $1 \leq j \leq m + 1$ denote by L_j the intersection

$$L_j = \bigcap_{i \neq j} K_i.$$

Our induction assumption for the case $|F| = m$ implies that each L_j is non-empty. Let M denote the convex hull of the union $L_1 \cup \cdots \cup L_{m+1}$.

If $x \in M$ then Carathéodory's Theorem 5.3.2 implies that x lies in the convex hull of the union $L_{i_1} \cup L_{i_2} \cup \cdots \cup L_{i_{n+1}}$ for some $1 \leq i_1 \leq \cdots \leq i_{n+1} \leq m + 1$. For each $i \notin \{i_1, \ldots, i_{n+1}\}$,

$$L_{i_1} \cup L_{i_2} \cup \cdots \cup L_{i_{n+1}} \subseteq K_i,$$

so that $x \in K_i$. In other words,

$$M \subseteq \bigcup_{i=1}^{m+1} K_i.$$

Let $M_i = K_i \cap M$ for $1 \leq i \leq m + 1$. Note that, for each j,

$$\bigcap_{i \neq j} M_i = \bigcap_{i \neq j} K_i \cap M = L_j \cap M = L_j \neq \emptyset.$$

Since $M_1 \cup \cdots \cup M_{m+1} = M$ is convex, it follows from Klee's Theorem 5.3.1 that

$$\bigcap_{i=1}^{m+1} M_i \neq \emptyset.$$

Hence,

$$\bigcap_{i=1}^{m+1} K_i \supseteq \bigcap_{i=1}^{m+1} M_i \neq \emptyset.$$

$\square$

Theorem 5.3.3 is in fact a generalization of Theorem 3.3.1. To see this, suppose that $S = \{s_1, \ldots, s_n\}$ is a finite set, and associate to each s_i a distinct point $x_i \in \mathbf{R}^{n-1}$, chosen so that the collection $\{x_1, \ldots, x_n\}$ is affinely independent. Let Δ denote the *geometric* simplex in $\mathbf{R}^{n-1}$ having vertices $\{x_1, \ldots, x_n\}$. Subsets of S now correspond to *faces* of the simplex Δ, and Theorem 3.3.1 is now a specialization of Theorem 5.3.3 to families of faces of the simplex Δ.

5.4 Lutwak's containment theorem

We now turn to a beautiful application of Helly's theorem to the question of containment of convex bodies. Given compact convex sets K and L with non-empty interiors, is there a simple condition that guarantees that some *translate* of K is a subset of L? It turns out that the answer to this question is determined by the relationship of K to the simplices in $\mathbf{R}^n$ that contain L.

For $K \in \mathcal{K}^n$ and $v \in \mathbf{R}^n$, denote by $K + v$ the set

$$K + v = \{x + v \, : \, x \in K\}.$$

In other words, $K + v$ denotes the translation of the set K by the vector v.

Theorem 5.4.1 (Lutwak's containment theorem) *Let $K, L \in \mathcal{K}^n$ with non-empty interiors. The following are equivalent.*

(i) *For every simplex Δ such that $L \subseteq \Delta$, there exists $v \in \mathbf{R}^n$ such that $K + v \subseteq \Delta$.*

(ii) *There exists $v_0 \in \mathbf{R}^n$ such that $K + v_0 \subseteq L$.*

In other words, if every simplex containing L also contains a translate of K, then L itself contains a translate of K.

Proof The implication (ii) $\Rightarrow$ (i) is obvious. We show that (i) $\Rightarrow$ (ii).

To begin, suppose first that L is a *polytope*, with facets $L_1, L_2, \ldots, L_m$ and corresponding facet (outward) unit normal vectors $u_1, u_2, \ldots, u_m$. Assume also that every selection of n distinct unit normals u_j is a linearly independent set. (Were this not the case, a small perturbation of L would make it so.)

For each facet L_i, let H_i denote the $(n-1)$-dimensional hyperplane in $\mathbf{R}^n$ containing L_i, and let H_i^+ denote the closed half-space bounded by H_i and containing the polytope L. Finally, denote by T_i the set of vectors $v \in \mathbf{R}^n$ such that $K + v \subset H_i^+$. Since each H_i^+ is a (convex) closed half-space and K is compact, it is clear that each T_i is a non-empty closed convex set.

The independence condition on the unit normals $\{u_i\}$ implies that, for each distinct selection $u_{i_1}, u_{i_2}, \ldots, u_{i_{n+1}}$ of $n+1$ unit normals, either the corresponding intersection

$$H_{i_1,i_2,\ldots,i_{n+1}} = \bigcap_{s=1}^{n+1} H_{i_s}^+ \tag{5.7}$$

contains a simplex $\Delta_{i_1,i_2,\ldots,i_{n+1}}$ such that $L \subseteq \Delta_{i_1,i_2,\ldots,i_{n+1}}$, or this intersection contains a translate of the ball αB of radius α, for all $\alpha > 0$. (This is the case in which the intersection (5.7) is unbounded.) In the first case, the hypothesis of the theorem implies the existence of a vector $v \in \mathbf{R}^n$ such that $K + v \subseteq \Delta_{i_1,i_2,\ldots,i_{n+1}}$. In the second case there also exists v such that $K + v$ lies in the intersection (5.7). In either case, there exists $v \in T_{i_1} \cap \cdots \cap T_{i_{n+1}}$. In other words, each collection of $n + 1$ sets T_i has a non-empty intersection. Helly's Theorem 5.3.3 then implies the existence of a vector v such that

$$v \in \bigcap_{i=1}^m T_i.$$

In other words, $K + v \subseteq H_i^+$ for $i = 1, \ldots, m$. Since $L = H_1^+ \cap \cdots \cap H_m^+$, it follows that $K + v \subseteq L$.

Now suppose that L is an arbitrary compact convex set. Let $\{P_i\}_{i=1}^\infty$ be a decreasing collection of polytopes such that $P_i \to L$ as $i \to \infty$, and such that each n of the facet normals to P_i are linearly independent. If Δ is a simplex containing P_i, then $L \subseteq P_i \subseteq \Delta$, so there exists a vector w such that $K + w \subseteq \Delta$. Since the theorem holds for the polytopes P_i, it then follows that there exists a vector v_i for each i, such that $K + v_i \subseteq P_i$. Since the P_i are decreasing (with respect to the relation of subset containment), the sequence $\{v_i\}$ is bounded and must contain a convergent subsequence. Assume then without loss of generality that $v_i \to v$. Since $P_i \to L$, it follows that $K + v \subseteq L$. $\square$

5.5 Cauchy's surface area formula

We conclude this chapter with the following interpretation of the surface area $S(K)$ of a convex body in $\mathbf{R}^n$, which will be of use to us in the sequel. If $K \in \mathcal{K}^n$ and V is an $(n-1)$-dimensional subspace of $\mathbf{R}^n$, denote by $K|V$ the orthogonal projection of K onto V. Let ω_{n-1} denote the $(n-1)$-dimensional volume of the unit ball B_{n-1} in $\mathbf{R}^{n-1}$.

Lemma 5.5.1 *For $v \in \mathbf{S}^{n-1}$,*

$$\int_{\mathbf{S}^{n-1}} |u \cdot v| \, \mathrm{d}u = 2\omega_{n-1}.$$

Proof Recall from elementary calculus that

$$\int_{\mathbf{S}^{n-1}} |u \cdot v| \, du \approx \sum_i |u_i \cdot v| S(A_i),$$

where $\mathbf{S}^{n-1}$ is partitioned into many small regions A_i, having area $S(A_i)$, and where $u_i \in A_i$ for each i. Let $\widetilde{A}_i$ denote the orthogonal projection of A_i onto the tangent hyperplane to $\mathbf{S}^{n-1}$ at the point u_i. Denote by $\widetilde{A}_i|v^\perp$ the orthogonal projection of $\widetilde{A}_i$ onto the hyperplane $v^\perp$; that is, into the disk B_{n-1} in $v^\perp$. Because $\widetilde{A}_i$ is a flat region lying inside a hyperplane with unit normal u_i, we have $S(\widetilde{A}_i|v^\perp) = |u \cdot v| S(\widetilde{A}_i)$.

Meanwhile, for A_i sufficiently small, we have $S(A_i) \approx S(\widetilde{A}_i)$ and $S(A_i|v^\perp) \approx S(\widetilde{A}_i|v^\perp)$. Therefore,

$$\int_{\mathbf{S}^{n-1}} |u \cdot v| \, du \approx \sum_i S(A_i|v^\perp).$$

Since the collection of sets $\{A_i|v^\perp\}$ covers the disk B_i twice, projecting from both of the directions v and $-v$ (that is, from both hemispheres of $\mathbf{S}^{n-1}$), we have

$$\int_{\mathbf{S}^{n-1}} |u \cdot v| \, du \approx 2\omega_{n-1},$$

where the similarities converge to equalities in the limit, as the mesh of the partition $\{A_i\}$ goes to zero.   □

Theorem 5.5.2 (Cauchy's surface area formula) *For all $K \in \mathcal{K}^n$,*

$$S(K) = \frac{1}{\omega_{n-1}} \int_{\mathbf{S}^{n-1}} \mu_{n-1}(K|u^\perp) \, du. \tag{5.8}$$

Here $\mathbf{S}^{n-1}$ denotes the $(n-1)$-dimensional sphere; i.e. the set of all unit vectors in $\mathbf{R}^n$. Note that $\mu_{n-1}(K|u^\perp)$ is just the $(n-1)$-dimensional volume of the projection of K onto the subspace $u^\perp$.

Proof Let P be a compact convex polytope with facet unit normal vectors $v_1, \ldots, v_m$ and corresponding facet surface areas $\alpha_1, \ldots, \alpha_m$. If a facet P_i of P has unit normal vector v_i and surface area (that is, $(n-1)$-volume) α_i, then the projection $P_i|u^\perp$ has $(n-1)$-volume

$$\mu_{n-1}(P_i|u^\perp) = \alpha_i |u \cdot v_i|.$$

For $u \in \mathbf{S}^{n-1}$, the $(n-1)$-volume $\mu_{n-1}(P|u^\perp)$ is computed by summing the $(n-1)$-volumes of the projections of the facets of P. This sum gives twice the desired value, since each point of $P|u^\perp$ is hit twice (from above and below) as we sum over projections of all facets of P. Hence, we have

$$
\begin{aligned}
\int_{\mathbf{S}^{n-1}} \mu_{n-1}(P|u^\perp)\, du &= \int_{\mathbf{S}^{n-1}} \frac{1}{2} \sum_{i=1}^{m} \alpha_i |u \cdot v_i|\, du \\
&= \sum_{i=1}^{m} \alpha_i \frac{1}{2} \int_{\mathbf{S}^{n-1}} |u \cdot v_i|\, du \\
&= \sum_{i=1}^{m} \alpha_i \omega_{n-1} \\
&= \omega_{n-1} S(P),
\end{aligned}
$$

where the third equality follows from Lemma 5.5.1. Since (5.8) then holds for any convex polytope P, the equation holds for any convex body K by continuity. $\qquad\square$

Recall that, for *parallelotopes* $P \in \mathrm{Par}(n)$,

$$
\mu_{n-1}(P) = \tfrac{1}{2} S(P).
$$

We are therefore motivated to define

$$
\mu_{n-1}(K) = \tfrac{1}{2} S(K),
$$

for all $K \in \mathcal{K}^n$. The equation (5.8) now becomes

$$
\mu_{n-1}(K) = \frac{1}{2\omega_{n-1}} \int_{\mathbf{S}^{n-1}} \mu_{n-1}(K|u^\perp)\, du. \tag{5.9}
$$

Since every line ℓ through the origin in $\mathbf{R}^n$ meets $\mathbf{S}^{n-1}$ at exactly two points, we can rewrite this result as an integral over the projective space $\mathrm{Gr}(n,1)$ (i.e. the set of all lines ℓ through the origin in $\mathbf{R}^n$) instead of integrating over the sphere. We then renormalize the resultant measure on $\mathrm{Gr}(n,1)$ so that the space $\mathrm{Gr}(n,1)$ has total measure 1. This measure is called the *Haar probability measure* on $\mathrm{Gr}(n,1)$. Integrating with respect to the Haar probability measure on $\mathrm{Gr}(n,1)$ we obtain

$$
\mu_{n-1}(K) = \alpha \int_{\mathrm{Gr}(n,1)} \mu_{n-1}(K|\ell^\perp)\, d\ell,
$$

where α is a constant independent of K. To compute α, set $K = B_n$, the unit ball in $\mathbf{R}^n$, to obtain

$$
\frac{n\omega_n}{2} = \mu_{n-1}(B_n) = \alpha \int_{\mathrm{Gr}(n,1)} \mu_{n-1}(B_n|\ell^\perp)\, d\ell = \alpha\omega_{n-1},
$$

since the total measure of the space $\mathrm{Gr}(n,1)$ is 1. It follows that

$$\alpha = \frac{n\omega_n}{2\omega_{n-1}}.$$

Denote by τ_n the rotation invariant measure on $\mathrm{Gr}(n,1)$ normalized so that

$$\tau_n(\mathrm{Gr}(n,1)) = \frac{n\omega_n}{2\omega_{n-1}}. \tag{5.10}$$

The equation (5.9) now becomes

$$\mu_{n-1}(K) = \int_{\mathrm{Gr}(n,1)} \mu_{n-1}(K|\ell^{\perp})\, \mathrm{d}\tau_n. \tag{5.11}$$

5.6 Notes

For a general reference to the theory of convex bodies, see the book by Schneider [85] and the survey on valuations by McMullen and Schneider [72]. The normalization for intrinsic volumes is due to McMullen [70]. Groemer's extension Theorem 5.1.1 originally appeared in [32], which contains a comprehensive presentation of extension theorems for valuations on various classes of compact convex sets.

For a treatment of the combinatorial theory of the Euler characteristic, see [79, 80]. See also [75] for the connection between the Euler characteristic and algebraic topology. Hadwiger presented numerous geometric applications of the Euler characteristic in [37, 39]. For a modern treatment of the Euler characteristic as a valuation on convex sets, see [71, 72, 85].

Helly's Theorem 5.3.3 first appeared in [43]; the proof given in this section is due to Hadwiger [37]. Theorem 5.3.2 is due to Carathéodory [85]. The proof of Carathéodory's theorem presented above may also be found in [37], where it appeared as a special case of a more general theorem of Steinitz [93]. Klee's Theorem 5.3.1 was noted by Klee in [56], and has many generalizations (see [6, 28, 37, 62]; also [18, pp. 123–128]). A non-numerical form (in terms of cone functions) was given by Chen [11]. A general survey on Helly's theorem, its proofs, variations and applications, may be found in [18] and in [21].

It is conceivable that the generalization of the Euler–Schläfli–Poincaré formula (5.2.4) to intrinsic volumes should lead to interesting quantitative generalizations of Klee's Theorem 5.3.1 and thereby of Helly's Theorem 5.3.3. Lutwak's containment Theorem 5.4.1 appeared in [68].

Cauchy published his surface area formula (Theorem 5.5.2) for dimensions 2 and 3 in [9] and [10]. Generalizations of Cauchy's formula are treated in Sections 7.4 and 9.4 (see also [82, p. 218] and [85, p. 295]).

6

Invariant measures on Grassmannians

Before we extend the notions introduced in Chapter 4 to the lattice of polyconvex sets we shall require a deeper understanding of the lattice of subspaces of $\mathbf{R}^n$. In Section 6.1 we introduce the the flag coefficients, which rely on a crucial choice of normalization for the rotation invariant measures on real Grassmannians. This construction leads in turn to continuous analogues of the extremal combinatorial results of Chapter 3, here in the context of subspaces. The flag coefficients will reappear in Chapters 8–10 as we investigate connections between the integral geometry of Grassmannians and the intrinsic volumes on polyconvex sets (see especially Section 9.4).

6.1 The lattice of subspaces

Let $\mathrm{Mod}(n)$ denote the set of all linear subspaces of $\mathbf{R}^n$; that is, the set of all linear varieties passing through the origin (having fixed an origin once and for all). The set $\mathrm{Mod}(n)$ is a partially ordered set under the relation of inclusion of linear subspaces. Moreover, it is a lattice, whereby the join $x \vee y$ and the meet $x \wedge y$ of two elements $x, y \in \mathrm{Mod}(n)$ are defined respectively as the linear subspaces spanned by x and y and as the intersection of x and y. The lattice $\mathrm{Mod}(n)$ may be viewed as a continuous analogue of the lattice $P(S)$ of subsets of a set S with n elements. Note, however, that this analogy is only a partial one, since the distributive law governing unions and intersections of subsets of S does not hold in the lattice $\mathrm{Mod}(n)$. Nonetheless, this analogy shall carry us as far as we need to go.

In the lattice $\mathrm{Mod}(n)$ one defines the notions of *chain*, *flag*, and *rank* as for the lattice of subsets of $\mathbf{R}^n$. In particular, an element x of $\mathrm{Mod}(n)$ has rank k (that is, $r(x) = k$), whenever x is a linear subspace of dimension

k. The subspace $\{0\}$ is the minimal element of the lattice $\mathrm{Mod}(n)$. Much as the group of permutations of the set S acts naturally on $P(S)$, the *orthogonal group* $O(n)$ (that is, the group of rotations about the origin and reflections across hyperplanes through the origin) acts naturally on the lattice $\mathrm{Mod}(n)$.

The set of all elements of $\mathrm{Mod}(n)$ of dimension (rank) k, denoted $\mathrm{Gr}(n,k)$, is called the *Grassmannian*. Our objective is to describe the invariant (Haar) measure that acts on $\mathrm{Gr}(n,k)$. It is known that this measure is unique up to a common factor (but we will not prove this). If our analogy is not misleading, the total measure of $\mathrm{Gr}(n,k)$ should be an analogue of the binomial coefficient.

To begin, consider the invariant measure τ_n on $\mathrm{Gr}(n,1)$; that is, on the set of all straight lines through the origin. Denote

$$[n] = \tau_n(\mathrm{Gr}(n,1)) = \frac{n\omega_n}{2\omega_{n-1}}, \tag{6.1}$$

where ω_n denotes the volume of the unit ball B_n in $\mathbf{R}^n$. The real numbers $[n]$ will play a role analogous to that of the positive integers n in the discrete case.

The measure τ_n may be viewed constructively in the following way. Let the invariant measure on the unit sphere $\mathbf{S}^{n-1}$ (that is, on the surface of the unit ball) be denoted by σ_{n-1}. For any measurable subset A of $\mathrm{Gr}(n,1)$, let A' be the subset of the unit sphere $\mathbf{S}^{n-1}$ defined by

$$A' = \bigcup_{x \in A} x \cap \mathbf{S}^{n-1}.$$

It then follows from (5.10) that

$$\tau_n(A) = \frac{\sigma_{n-1}(A')}{2\omega_{n-1}}.$$

To check this, note that the surface area of the unit sphere is

$$\sigma_{n-1}(\mathbf{S}^{n-1}) = n\omega_n$$

so that

$$\tau_n(\mathrm{Gr}(n,1)) = \frac{n\omega_n}{2\omega_{n-1}}.$$

Again it is clear that the measure τ_n is invariant under rotations.

Let $\mathrm{Flag}(n)$ be the set of flags in $\mathrm{Mod}(n)$. For $x \in \mathrm{Mod}(n)$, denote by $\mathrm{Flag}(x)$ the set of all flags that contain x; that is, the set of all sequences $(x_0, x_1, \ldots, x_n)$ of $x_i \in \mathrm{Mod}(n)$, where $\dim(x_i) = i$, such that

$x_0 \leq x_1 \leq \cdots \leq x_n$, and such that one of the x_i equals x. Note that $x_0 = \{0\}$ and $x_n = \mathbf{R}^n$.

For fixed x_k, the set of all sequences $(x_k, x_{k+1}, \ldots, x_n)$ with $x_i \in$ Mod(n), such that dim$(x_i) = i$ and $x_i \leq x_{i+1}$, is isomorphic to Flag$(n - k)$. Similarly, the set of sequences $(x_0, x_1, \ldots, x_k)$ is isomorphic to Flag(k).

Denote by ϕ_n the invariant measure on the set Flag(n), to be computed as follows. The measure τ_n on Gr$(n, 1)$ induces a measure $\tilde{\tau}_n$ on the set Gr$(n, n - 1)$ via orthogonal duality. If $f(x_0, x_1, \ldots, x_n)$ is a simple function on the set Flag(n), then let

$$\int f \, d\phi_n = \int \int f(x_0, x_1, \ldots, x_n) \, d\phi_{n-1}(x_0, \ldots, x_{n-1}) \, d\tilde{\tau}_n,$$

so that the measure ϕ_n is defined inductively. It is clearly invariant.

A more explicit form for ϕ_n can be obtained by the following argument. If $(x_0, x_1, \ldots, x_n)$ is a flag in Mod(n), let

$$y_1 = x_1, \ y_2 = x_1^{\perp} \cap x_2, \ y_1 = x_2^{\perp} \cap x_3, \ldots, y_n = x_{n-1}^{\perp} \cap x_n.$$

The sequence $(y_1, y_2, \ldots, y_n)$ is a sequence of orthogonal straight lines, a frame. Conversely, given a frame $(y_1, y_2, \ldots, y_n)$, we obtain a flag by setting

$$x_0 = \{0\}, \ x_1 = y_1, \ x_2 = y_1 \vee y_2, \ x_3 = y_1 \vee y_2 \vee y_3, \ldots$$

This defines a one-to-one correspondence between flags and frames. If $f(x_0, x_1, \ldots, x_n)$ is a real-valued (measurable) function on flags, let $\bar{f}(y_1, y_2, \ldots, y_n)$ be the corresponding function on frames. Then

$$\int f \, d\phi_n = \int \int \cdots \int \bar{f}(y_1, y_2, \ldots, y_n) \, d\tau_1(y_n) \, d\tau_2(y_{n-1}) \cdots d\tau_n(y_1).$$

This result can be read in the simpler language of combinatorics. Once the line y_1 has been chosen, which can be done in τ_n ways, the subspace x_2 is determined by the choice of a straight line y_2 in the space orthogonal to x_1 (through the origin), which can be done in τ_{n-1} ways, etc.

The measure of Flag(n) turns out to be

$$\phi_n(\text{Flag}(n)) = [n][n - 1] \ldots [2][1],$$

also denoted $[n]!$, where $[1] = 1$. Note that

$$[n]! = \frac{n! \omega_n \omega_{n-1} \cdots \omega_1}{2^n \omega_{n-1} \omega_{n-2} \cdots \omega_0} = \frac{n! \omega_n}{2^n}. \tag{6.2}$$

We now define an invariant measure on $\mathrm{Gr}(n,k)$. For $A \subseteq \mathrm{Gr}(n,k)$, let $\mathrm{Flag}(A)$ be the set of all flags $(x_0, x_1, \ldots, x_n)$ such that $x_k \in A$. Set

$$\nu_k^n(A) = \frac{1}{[k]![n-k]!}\phi_n(\mathrm{Flag}(A)). \qquad (6.3)$$

To justify this normalization combinatorially, note that, for each $x_k \in A$, there are exactly $[k]![n-k]!$ flags containing x_k; that is, to choose a flag containing x_k one must choose a frame for the vector space x_k, of which there are $[k]!$ choices, and a frame for the $(n-k)$-dimensional complementary space $x_k^\perp$, of which there are $[n-k]!$ choices.

The measure ν_k^n is evidently invariant under rotations, and we have

$$\nu_k^n(\mathrm{Gr}(n,k)) = \frac{[n]!}{[k]![n-k]!} = \begin{bmatrix} n \\ k \end{bmatrix}.$$

These values, called *flag coefficients*, are continuous analogues of the binomial coefficients of the discrete lattice in Chapter 3. From (6.2) we obtain

$$\begin{bmatrix} n \\ k \end{bmatrix} = \frac{n!}{k!(n-k)!}\frac{\omega_n}{\omega_k\omega_{n-k}} = \binom{n}{k}\frac{\omega_n}{\omega_k\omega_{n-k}}. \qquad (6.4)$$

6.2 Computing the flag coefficients

It is not difficult to compute the numerical values of the numbers $[n]$ and their associated flag coefficients. We begin by deriving the following well known formula for the volume ω_n of the unit ball in $\mathbf{R}^n$.

Proposition 6.2.1 *For $n \geq 1$*

$$\omega_n = \frac{\pi^{n/2}}{\Gamma((n/2)+1)}.$$

Here $\Gamma(t)$ denotes the Euler gamma function, given by

$$\Gamma(t) = \int_0^\infty \mathrm{e}^{-x}x^{t-1}\mathrm{d}x.$$

Proof We first compute the integral

$$\int_{-\infty}^\infty \mathrm{e}^{-x^2}\,\mathrm{d}x.$$

We can think of the square of this integral as the product of integrals in two independent variables, x and y.

$$\left(\int_{-\infty}^{\infty} e^{-x^2}\, dx\right)^2 = \int_{-\infty}^{\infty} e^{-x^2}\, dx \int_{-\infty}^{\infty} e^{-y^2}\, dy = \int_{-\infty}^{\infty} e^{-(x^2+y^2)}\, dx\, dy$$

On switching to polar coordinates in the xy–plane, we obtain

$$\begin{aligned}
\left(\int_{-\infty}^{\infty} e^{-x^2}\, dx\right)^2 &= \int_{-\infty}^{\infty} e^{-(x^2+y^2)}\, dx\, dy \\
&= \int_{0}^{2\pi} \int_{0}^{\infty} e^{-r^2} r\, dr\, d\theta \\
&= 2\pi \int_{0}^{\infty} e^{-r^2} r\, dr \\
&= 2\pi \left(\frac{-e^{-r^2}}{2}\right)_{0}^{\infty} \\
&= \pi,
\end{aligned}$$

so that

$$\int_{-\infty}^{\infty} e^{-x^2}\, dx = \sqrt{\pi}. \tag{6.5}$$

Following a similar argument in n-variables,

$$\begin{aligned}
\pi^{n/2} &= \left(\int_{-\infty}^{\infty} e^{-x^2}\, dx\right)^n \\
&= \int_{-\infty}^{\infty} e^{-x_1^2}\, dx_1 \cdots \int_{-\infty}^{\infty} e^{-x_n^2}\, dx_n \\
&= \int_{-\infty}^{\infty} e^{-(x_1^2+\cdots+x_n^2)}\, dx_1 \cdots dx_n \\
&= \int_{u \in S^{n-1}} \int_{0}^{\infty} e^{-r^2} r^{n-1}\, dr\, du \\
&= n\omega_n \int_{0}^{\infty} e^{-r^2} r^{n-1}\, dr,
\end{aligned}$$

since the surface area of the sphere $\mathbf{S}^{n-1}$ is equal to $n\omega_n$. On substituting $y = r^2$, we obtain

$$\begin{aligned}
\pi^{n/2} &= n\omega_n \int_{0}^{\infty} e^{-r^2} r^{n-1}\, dr, \\
&= \frac{n}{2}\omega_n \int_{0}^{\infty} e^{-y} y^{(n/2)-1}\, dy,
\end{aligned}$$

$$= \frac{n}{2}\omega_n\Gamma\left(\frac{n}{2}\right)$$

$$= \omega_n\Gamma\left(\frac{n}{2}+1\right)$$

Hence, we have

$$\omega_n = \frac{\pi^{n/2}}{\Gamma(\frac{n}{2}+1)}.$$

$\square$

Proposition 6.2.1 can be given a more computationally pliable form, provided that we examine separately the cases in which the dimension n of the unit ball is even or odd.

Proposition 6.2.2 *Let k be a non-negative integer. Then*

$$\omega_{2k} = \frac{\pi^k}{k!} \quad and \quad \omega_{2k+1} = \frac{2^{2k+1}\pi^k k!}{(2k+1)!}$$

Proof It immediately follows from Proposition 6.2.1 that

$$\omega_{2k} = \frac{\pi^k}{\Gamma(k+1)} = \frac{\pi^k}{k!}.$$

To compute ω_{2k+1} first note that

$$\Gamma(1/2) = \int_0^\infty e^{-x} x^{-\frac{1}{2}} \, dx.$$

On substituting $x = y^2$ and $dx = 2y \, dy$ we obtain

$$\Gamma(1/2) = \int_0^\infty e^{-y^2} \frac{1}{y} 2y \, dy = \int_{-\infty}^\infty e^{-y^2} \, dy = \sqrt{\pi},$$

where the last equality follows from (6.5). It now follows from Proposition 6.2.1 that

$$\begin{aligned}
\omega_{2k+1} &= \frac{\pi^{\frac{2k+1}{2}}}{\Gamma(\frac{2k+1}{2}+1)} \\
&= \frac{\pi^k \sqrt{\pi}}{\frac{2k+1}{2}\frac{2k-1}{2}\cdots\frac{1}{2}\sqrt{\pi}} \\
&= \frac{2^{k+1}\pi^k}{(2k+1)(2k-1)\cdots(3)(1)}
\end{aligned}$$

$$= \frac{2^{k+1}\pi^k 2^k k!}{(2k+1)!}$$

$$= \frac{2^{2k+1}\pi^k k!}{(2k+1)!}$$

$\square$

Proposition 6.2.2 leads to easy computations of $[n]$. Once again there are two formulas, depending on the parity of n.

Proposition 6.2.3 *Let k be a positive integer. Then*

$$[2k] = \frac{2\pi(2k-1)!}{4^k(k-1)!(k-1)!} = \frac{2\pi k}{4^k}\binom{2k-1}{k},$$

and

$$[2k+1] = \frac{4^k(k!)^2}{(2k)!} = 4^k\binom{2k}{k}^{-1}.$$

Proof Computing directly from the definition (6.1) and Proposition 6.2.2 we get

$$[2k] = \frac{2k\omega_{2k}}{2\omega_{2k-1}}$$

$$= \frac{k\pi^k}{k!}\frac{(2k-1)!}{2^{2k-1}\pi^{k-1}(k-1)!}$$

$$= \frac{2\pi k}{4^k}\binom{2k-1}{k}.$$

Similarly,

$$[2k+1] = \frac{(2k+1)\omega_{2k+1}}{2\omega_{2k}}$$

$$= \frac{(2k+1)2^{2k+1}\pi^k k!}{(2k+1)!}\frac{k!}{2\pi^k}$$

$$= \frac{4^k k! k!}{(2k)!}.$$

$\square$

The first few values of ω_n, $[n]$, and $[n]!$ run as follows:

n	$\dfrac{\pi^{\frac{n}{2}}}{\Gamma(\frac{n}{2}+1)}$	$\dfrac{n\omega_n}{2\omega_{n-1}}$	$\dfrac{n!\omega_n}{2^n}$
n	ω_n	$[n]$	$[n]!$
0	1	1	1
1	2	1	1
2	π	$\frac{\pi}{2}$	$\frac{\pi}{2}$
3	$\frac{4\pi}{3}$	2	π
4	$\frac{\pi^2}{2}$	$\frac{3\pi}{4}$	$\frac{3\pi^2}{4}$
5	$\frac{8\pi^2}{15}$	$\frac{8}{3}$	$2\pi^2$
6	$\frac{\pi^3}{6}$	$\frac{15\pi}{16}$	$\frac{15\pi^3}{8}$
7	$\frac{16\pi^3}{105}$	$\frac{16}{5}$	$6\pi^3$
8	$\frac{\pi^4}{24}$	$\frac{35\pi}{32}$	$\frac{105\pi^4}{16}$
9	$\frac{32\pi^4}{945}$	$\frac{128}{35}$	$24\pi^4$
10	$\frac{\pi^5}{120}$	$\frac{315\pi}{256}$	$\frac{945\pi^5}{32}$

$$(6.6)$$

Proposition 6.2.3 leads to a number of interesting relations among the numbers $[n]$. For example, one can easily show that, for $k > 0$,

$$[2k][2k+1] = k\pi,$$

and

$$[2k-1][2k] = \frac{(2k-1)\pi}{2},$$

from which it follows that

$$\frac{[n+2]}{[n]} = \frac{n+1}{n},$$

for all $n > 0$.

Using either of Propositions 6.2.2 and 6.2.3 one may easily compute the values of the flag coefficients. Once again the formulas vary according to the parity of the parameters.

Proposition 6.2.4 *Let m and k be positive integers. Then*

$$\begin{bmatrix} 2m \\ 2k \end{bmatrix} = \begin{pmatrix} 2m \\ 2k \end{pmatrix}\begin{pmatrix} m \\ k \end{pmatrix}^{-1} \quad and$$

$$\begin{bmatrix} 2m \\ 2k+1 \end{bmatrix} = \frac{\pi}{4^m}\frac{(2m)!}{k!m!(m-k-1)!} = \frac{2m\pi}{4^m}\begin{pmatrix} 2m-1 \\ k,m,m-k-1 \end{pmatrix},$$

while

$$\begin{bmatrix} 2m+1 \\ 2k \end{bmatrix} = 4^k \binom{m}{k} \binom{2k}{k}^{-1} \quad and$$

$$\begin{bmatrix} 2m+1 \\ 2k+1 \end{bmatrix} = 4^{m-k} \binom{m}{m-k} \binom{2(m-k)}{m-k}^{-1}.$$

Proof Combining (6.4) with the formulas of Proposition 6.2.2, we obtain

$$\begin{bmatrix} 2m \\ 2k \end{bmatrix} = \binom{2m}{2k} \frac{\omega_{2m}}{\omega_{2k}\omega_{2m-2k}} = \binom{2m}{2k} \frac{\pi^m}{m!} \frac{k!(m-k)!}{\pi^k \pi^{m-k}} = \binom{2m}{2k} \binom{m}{k}^{-1}.$$

With a little more effort, we have

$$\begin{aligned}
\begin{bmatrix} 2m \\ 2k+1 \end{bmatrix} &= \binom{2m}{2k+1} \frac{\omega_{2m}}{\omega_{2k+1}\omega_{2m-2k-1}} \\
&= \binom{2m}{2k+1} \frac{\pi^m}{m!} \frac{(2k+1)!}{2^{2k+1}\pi^k k!} \frac{(2m-2k-1)!}{2^{2m-2k-1}\pi^{m-k-1}(m-k-1)!} \\
&= \binom{2m}{2k+1} \frac{\pi}{m!2^{2m}} \frac{(2k+1)!(2m-2k-1)!}{k!(m-k-1)!} \\
&= \frac{\pi}{4^m} \frac{(2m)!}{k!m!(m-k-1)!}.
\end{aligned}$$

Computing once again,

$$\begin{aligned}
\begin{bmatrix} 2m+1 \\ 2k \end{bmatrix} &= \binom{2m+1}{2k} \frac{\omega_{2m+1}}{\omega_{2k}\omega_{2m-2k+1}} \\
&= \binom{2m+1}{2k} \frac{2^{2m+1}\pi^m m!}{(2m+1)!} \frac{k!}{\pi^k} \frac{(2m-2k+1)!}{2^{2m-2k+1}\pi^{m-k}(m-k)!} \\
&= \frac{4^k m!k!}{(2k)!(m-k)!} \frac{k!}{k!} \\
&= 4^k \binom{m}{k} \binom{2k}{k}^{-1}.
\end{aligned}$$

The final formula now follows easily from the dual symmetry of the flag coefficients:

$$\begin{bmatrix} 2m+1 \\ 2k+1 \end{bmatrix} = \begin{bmatrix} 2m+1 \\ 2(m-k) \end{bmatrix} = 4^{m-k} \binom{m}{m-k} \binom{2(m-k)}{m-k}^{-1},$$

by the previous computation. $\qquad\qquad\qquad\qquad\qquad\qquad\qquad\qquad\square$

By means of Proposition 6.2.4 one may easily generate the following Pascal triangle for the flag coefficients $\begin{bmatrix} n \\ k \end{bmatrix}$, as $n = 1, \dots, 8$.

$$
\begin{array}{ccccccccccccccc}
&&&&&&& 1 && 1 &&&&& \\
&&&&&& 1 && \frac{\pi}{2} && 1 &&&& \\
&&&&& 1 && 2 && 2 && 1 &&& \\
&&&& 1 && \frac{3\pi}{4} && 3 && \frac{3\pi}{4} && 1 && \\
&&& 1 && \frac{8}{3} && 4 && 4 && \frac{8}{3} && 1 & \\
&& 1 && \frac{15\pi}{16} && 5 && \frac{15\pi}{8} && 5 && \frac{15\pi}{16} && 1 \\
& 1 && \frac{16}{5} && 6 && 8 && 8 && 6 && \frac{16}{5} && 1 \\
1 && \frac{35\pi}{32} && 7 && \frac{105\pi}{32} && \frac{35}{3} && \frac{105\pi}{32} && 7 && \frac{35\pi}{32} && 1
\end{array}
$$

$$(6.7)$$

Certain patterns are immediately noticeable. For example, the table (6.7) suggests that

$$\begin{bmatrix} n \\ 2 \end{bmatrix} = n - 1$$

for all $n \geq 2$. This indeed follows easily from Proposition 6.2.4. It is also not difficult to verify that

$$\begin{bmatrix} n \\ \frac{n-1}{2} \end{bmatrix} = 2^{\frac{n-1}{2}},$$

for all *odd* positive integers n.

If n is *even*, the value of the middle flag coefficient is given by one of two formulas, depending on the congruence class of n modulo 4. Computations similar to those preceding show that if $n \equiv 0 \bmod 4$ then

$$\begin{bmatrix} n \\ \frac{n}{2} \end{bmatrix} = \binom{n}{\frac{n}{2}} \binom{\frac{n}{2}}{\frac{n}{4}}^{-1}.$$

If $n \equiv 2 \bmod 4$ then

$$\begin{bmatrix} n \\ \frac{n}{2} \end{bmatrix} = \binom{n}{\frac{n}{2}} \binom{\frac{n}{2}}{\frac{n-2}{4}} \frac{(n+2)\pi}{2^{n+2}} = \binom{n}{\frac{n}{2}, \frac{n-2}{4}, \frac{n+2}{4}} \frac{(n+2)\pi}{2^{n+2}}.$$

Recall that, for fixed n, the classical binomial coefficient $\binom{n}{k}$ is maximized at $k = n/2$ if n is even, or $k = (n-1)/2$ if n is odd. In the next section we will show that flag coefficients are also maximized 'in the middle,' as the table (6.7) suggests.

6.3 Properties of the flag coefficients

The flag coefficients satisfy a number of properties analogous to those of the binomial coefficients. For example,

$$\begin{bmatrix} n \\ k \end{bmatrix} = \begin{bmatrix} n \\ n-k \end{bmatrix}.$$

In analogy to Pascal's triangle relations we have the following.

Proposition 6.3.1 *For* $1 \leq k \leq n-1$,

$$\frac{\omega_{k-1}\omega_{n-k}}{\omega_{n-1}} \begin{bmatrix} n-1 \\ k-1 \end{bmatrix} + \frac{\omega_k \omega_{n-k-1}}{\omega_{n-1}} \begin{bmatrix} n-1 \\ k \end{bmatrix} = \frac{\omega_k \omega_{n-k}}{\omega_n} \begin{bmatrix} n \\ k \end{bmatrix} \qquad (6.8)$$

Proof We prove (6.8) by direct computation:

$$\begin{aligned}
\frac{\omega_{k-1}\omega_{n-k}}{\omega_{n-1}} \begin{bmatrix} n-1 \\ k-1 \end{bmatrix} + \frac{\omega_k \omega_{n-k-1}}{\omega_{n-1}} \begin{bmatrix} n-1 \\ k \end{bmatrix} &= \binom{n-1}{k-1} + \binom{n-1}{k} \\
&= \binom{n}{k} \\
&= \frac{\omega_k \omega_{n-k}}{\omega_n} \begin{bmatrix} n \\ k \end{bmatrix}.
\end{aligned}$$

□

In the discrete case, Pascal's triangle for binomial coefficients is much cleaner:

$$\binom{n-1}{k-1} + \binom{n-1}{k} = \binom{n}{k}.$$

A similar expression exists for the flag coefficients, with a relation of inequality. The inequality will follow from a more elementary additive inequality for the numbers $[n]$.

Recall from elementary calculus that the volume of the unit n-ball may be computed by integrating over its $(n-1)$-dimensional slices; that is,

$$\omega_n = 2 \int_0^1 \omega_{n-1}(1-x^2)^{\frac{n-1}{2}} \, dx.$$

The substitutions $x = \sin\theta$ and $dx = \cos\theta \, d\theta$ yield

$$\omega_n = 2\omega_{n-1} \int_0^{\frac{\pi}{2}} \cos^n \theta \, d\theta.$$

In other words,

$$[n] = \frac{n\omega_n}{2\omega_{n-1}} = n \int_0^{\frac{\pi}{2}} \cos^n \theta \, d\theta. \qquad (6.9)$$

This equation leads in turn to the following proposition.

Proposition 6.3.2 *For all positive integers m and n,*

$$[n] + [m] > [n + m].$$

Proof If $0 \le \theta \le \pi/2$, then $0 \le \cos \theta \le 1$. From (6.9) we then obtain

$$
\begin{aligned}
[n] + [m] &= n \int_0^{\frac{\pi}{2}} \cos^n \theta \, d\theta + m \int_0^{\frac{\pi}{2}} \cos^m \theta \, d\theta \\
&> n \int_0^{\frac{\pi}{2}} \cos^{n+m} \theta \, d\theta + m \int_0^{\frac{\pi}{2}} \cos^{n+m} \theta \, d\theta \\
&= (n + m) \int_0^{\frac{\pi}{2}} \cos^{n+m} \theta \, d\theta = [n + m].
\end{aligned}
$$

$\square$

Proposition 6.3.2 and Theorem 6.3.3 illustrate a fundamental difference between the combinatorics of finite sets and the measure structure of this vector space analogue.

Theorem 6.3.3 (The Pascal inequality for flag coefficients) *For $1 \le k \le n - 1$,*

$$\begin{bmatrix} n-1 \\ k-1 \end{bmatrix} + \begin{bmatrix} n-1 \\ k \end{bmatrix} > \begin{bmatrix} n \\ k \end{bmatrix}.$$

Proof For $1 \le k \le n - 1$, we have

$$
\begin{aligned}
\begin{bmatrix} n-1 \\ k-1 \end{bmatrix} + \begin{bmatrix} n-1 \\ k \end{bmatrix} &= \frac{[n-1]!}{[k-1]![n-k]!} + \frac{[n-1]!}{[k]![n-k-1]!} \\
&= \frac{[n-1]!([k] + [n-k])}{[k]![n-k]!}.
\end{aligned}
$$

By Proposition 6.3.2, $[k] + [n - k] > [k + (n - k)] = [n]$, so that

$$\begin{bmatrix} n-1 \\ k-1 \end{bmatrix} + \begin{bmatrix} n-1 \\ k \end{bmatrix} > \frac{[n-1]![n]}{[k]![n-k]!} = \frac{[n]!}{[k]![n-k]!}.$$

$\square$

Recall that the real numbers $[n]$ play a role analogous to that of the positive integers n in the discrete case. The following proposition demonstrates in part the expedience of our choice of normalization for the measure τ_n; that is, for the value of $[n]$.

Proposition 6.3.4 *The map* $n \longmapsto [n]$ *is an increasing function.*

Proof From Proposition 6.2.1 we have

$$[n] = \frac{n\omega_n}{2\omega_{n-1}} = \frac{n\sqrt{\pi}}{2}\frac{\Gamma\left(\frac{n+1}{2}\right)}{\Gamma\left(\frac{n+2}{2}\right)} = \frac{n\sqrt{\pi}}{2}\frac{\Gamma\left(\frac{n+1}{2}\right)}{\frac{n}{2}\Gamma\left(\frac{n}{2}\right)} = \sqrt{\pi}\frac{\Gamma\left(\frac{n+1}{2}\right)}{\Gamma\left(\frac{n}{2}\right)}.$$

Define a function f on the positive real numbers by

$$f(t) = \sqrt{\pi}\frac{\Gamma\left(t+\frac{1}{2}\right)}{\Gamma(t)}.$$

Since $f(\frac{n}{2}) = [n]$, it is sufficient to show that f is an increasing function of t. Recall that [4, p. 15]

$$\Gamma(t) = \lim_{k\to\infty}\frac{k^t k!}{t(t+1)\cdots(t+k)}.$$

This implies that

$$
\begin{aligned}
f(t) &= \sqrt{\pi}\left(\lim_{k\to\infty}\frac{k^{t+(1/2)}k!}{(t+\frac{1}{2})(t+\frac{1}{2}+1)\cdots(t+\frac{1}{2}+k)}\right) \times \\
&\qquad \left(\lim_{k\to\infty}\frac{k^t k!}{t(t+1)\cdots(t+k)}\right)^{-1} \\
&= \sqrt{\pi}\lim_{k\to\infty}\frac{t(t+1)\cdots(t+k)\sqrt{k}}{(t+\frac{1}{2})(t+\frac{1}{2}+1)\cdots(t+\frac{1}{2}+k)} \qquad (6.10)
\end{aligned}
$$

Since the function

$$\frac{t}{t+\frac{1}{2}} = 1 - \frac{1}{2t+1}$$

is increasing with respect to $t > 0$, so is the product on the right-hand side of (6.10). It follows that f is increasing for $t > 0$, and we conclude that $[n]$ is an increasing function of the positive integers. □

From Proposition 6.3.4 we deduce the following useful property for the generalized factorial $[n]!$:

Proposition 6.3.5 *For* $1 \le k \le l \le n/2$,

$$[k]![n-k]! \ge [l]![n-l]!$$

Proof To begin, note that

$$0 \leq k \leq l \leq \frac{n}{2} \leq n - l \leq n - k \leq n.$$

It follows from Proposition 6.3.4 that

$$[n - k] \cdots [n - l + 1] \geq [l] \cdots [k + 1].$$

(Note that there are $l - k$ factors on each side of this inequality.) Multiplying on both sides, we obtain

$$[n - k]![k]! \geq [n - l]![l]!$$

$\square$

The flag coefficients satisfy the following property, in evident analogy to the classical binomial coefficients.

Proposition 6.3.6 *For* $1 \leq k \leq n$,

$$\begin{bmatrix} n \\ k \end{bmatrix} \leq \begin{bmatrix} n \\ \langle n/2 \rangle \end{bmatrix}.$$

Once again $\langle n/2 \rangle$ denotes the greatest integer less than or equal to $n/2$.

Proof Since

$$\begin{bmatrix} n \\ k \end{bmatrix} = \frac{[n]!}{[k]![n - k]!} = \begin{bmatrix} n \\ n - k \end{bmatrix},$$

it is sufficient to consider the case in which $k < \langle n/2 \rangle$. On applying Proposition 6.3.5 to the case $l = \langle n/2 \rangle$, we find that

$$[n - k]![k]! \geq [\langle n/2 \rangle]![n - \langle n/2 \rangle]!,$$

so that

$$\frac{[n]!}{[n - k]![k]!} \leq \frac{[n]!}{[\langle n/2 \rangle]![n - \langle n/2 \rangle]!}.$$

$\square$

6.4 A continuous analogue of Sperner's theorem

Define a measure ν_n on $\mathrm{Mod}(n)$ by taking the direct sum of the measures ν_k^n. That is, for any measurable subset $A \subseteq \mathrm{Mod}(n)$, define

$$\nu_n(A) = \sum_{k=0}^{n} \nu_k^n(A \cap \mathrm{Gr}(n, k)).$$

The measure ν_n satisfies the following analogue of the classical L.Y.M. inequality (3.1).

Theorem 6.4.1 (The continuous L.Y.M. inequality) *Let $A \subseteq$* Mod(n) *be an antichain. For $0 \leq k \leq n$ let*

$$A_k = A \cap \mathrm{Gr}(n,k),$$

so that

$$A = \bigcup_k A_k$$

is a disjoint union. Then

$$\sum_{k=0}^{n} \frac{\nu_k^n(A_k)}{\genfrac{[}{]}{0pt}{}{n}{k}} \leq 1. \qquad (6.11)$$

Proof For each $0 \leq k \leq n$, the measure of flags meeting A_k is given by

$$\phi_n(\mathrm{Flag}(A_k)) = \nu_k^n(A_k)[k]![n-k]!$$

by the definition (6.3) of ν_k^n. Since every flag in Flag(n) meets A in at most one point, we have

$$\sum_{k=0}^{n} \nu_k^n(A_k)[k]![n-k]! = \sum_{k=0}^{n} \phi_n(\mathrm{Flag}(A_k)) = \phi_n(\mathrm{Flag}(A)) \leq [n]!$$

It follows that

$$\sum_{k=0}^{n} \frac{\nu_k^n(A_k)}{\genfrac{[}{]}{0pt}{}{n}{k}} \leq 1.$$

$\square$

We now have the necessary tools to prove a continuous analogue of Theorem 3.1.1.

Theorem 6.4.2 (The continuous Sperner theorem) *Let A be an antichain in* Mod(n). *Then*

$$\nu_n(A) \leq \begin{bmatrix} n \\ \langle n/2 \rangle \end{bmatrix}.$$

Evidently equality is attained in Theorem 6.4.2 when $A = \mathrm{Gr}(n, \langle n/2 \rangle)$. *Proof* We reason in analogy to the proof of Theorem 3.1.1. For $0 \leq k \leq n$ let $A_k = A \cap \mathrm{Gr}(n,k)$. Combining (6.11) and Proposition 6.3.6 we obtain

$$\sum_{k=0}^{n} \frac{\nu_k^n(A_k)}{\genfrac{[}{]}{0pt}{}{n}{\langle n/2 \rangle}} \leq \sum_{k=0}^{n} \frac{\nu_k^n(A_k)}{\genfrac{[}{]}{0pt}{}{n}{k}} \leq 1,$$

so that

$$\nu_n(A) = \sum_{k=0}^{n} \nu_k^n(A_k) \leq \left[\begin{matrix} n \\ \langle n/2 \rangle \end{matrix} \right].$$

□

In analogy to the discrete lattice $P(S)$, a subset $F \subseteq \mathrm{Mod}(n)$ is called an *r-family* if chains in F contain no more than r elements. Given an r-family F in $\mathrm{Mod}(n)$ let $F_k = F \cap \mathrm{Gr}(n, k)$. Since every flag in $\mathrm{Mod}(n)$ meets F in at most r elements, we have

$$\sum_{k=0}^{n} \phi_n(\mathrm{Flag}(F_k)) \leq [n]! \cdot r.$$

From (6.3) we then obtain the following generalization of (6.11):

$$\sum_{k=0}^{n} \frac{\nu_k^n(F_k)}{\left[\begin{smallmatrix} n \\ k \end{smallmatrix} \right]} \leq r. \tag{6.12}$$

This inequality leads in turn to a continuous analogue of Sperner's theorem for r-families (Theorem 3.1.2).

Theorem 6.4.3 *Let F be an r-family in* $\mathrm{Mod}(n)$. *Then*

$$\nu_n(F) \leq \left[\begin{matrix} n \\ \langle \frac{n+1}{2} \rangle \end{matrix} \right] + \left[\begin{matrix} n \\ \langle \frac{n+2}{2} \rangle \end{matrix} \right] + \cdots + \left[\begin{matrix} n \\ \langle \frac{n+r}{2} \rangle \end{matrix} \right].$$

Proof As in the proof of the discrete case (Theorem 3.1.2), relabel the flag coefficients $c_0, c_1, \ldots, c_n$ in *descending* order, so that $c_0 \geq c_1 \geq \cdots \geq c_n$; then relabel the numerators $\nu_k^n(F_k)$ in (6.12) by $x_0, x_1, \ldots, x_n$ so that each x_k is the numerator of that term of (6.12) having c_k as denominator. The inequality (6.12) now becomes

$$\sum_{k=0}^{n} \frac{x_k}{c_k} \leq r.$$

It then follows from Lemma 3.1.3 that

$$x_0 + x_1 + \cdots + x_n \leq c_0 + c_1 + \cdots + c_{r-1}.$$

In other words,

$$\nu_n(F) = \sum_{k=0}^{n} \nu_k^n(F_k) \leq \left[\begin{matrix} n \\ \langle \frac{n+1}{2} \rangle \end{matrix} \right] + \left[\begin{matrix} n \\ \langle \frac{n+2}{2} \rangle \end{matrix} \right] + \cdots + \left[\begin{matrix} n \\ \langle \frac{n+r}{2} \rangle \end{matrix} \right].$$

□

We now consider a continuous analogue of a question first posed by Sperner (see also Theorem 3.2.7). Given $A \subseteq \mathrm{Gr}(n, k)$, let $[A]_l$ denote the collection

$$[A]_l = \{W \in \mathrm{Gr}(n, l) : W \subseteq V \text{ for some } V \in A\}.$$

Here we assume that $0 \leq l \leq k \leq n$. Sperner's question, recast in the language of subspaces, asks for a lower bound on the measure of $[A]_l$, given only the measure $\nu_n(A)$. The continuous L.Y.M. inequality gives us such a bound.

Theorem 6.4.4 *For $A \subseteq \mathrm{Gr}(n, k)$,*

$$\nu_n([A]_l) \geq \frac{[k]![n-k]!}{[l]![n-l]!} \nu_n(A).$$

Proof Let $B_l = \mathrm{Gr}(n, l) - [A]_l$. If $W \in B_l$ then W cannot be contained in any $V \in A$. In other words, the set $A \cup B_l$ is an antichain. It then follows from (6.11) that

$$\frac{\nu_n(A)}{\begin{bmatrix} n \\ k \end{bmatrix}} + \frac{\nu_n(B_l)}{\begin{bmatrix} n \\ l \end{bmatrix}} \leq 1.$$

Meanwhile,

$$\nu_n(B_l) = \nu_n(\mathrm{Gr}(n, l)) - \nu_n([A]_l) = \begin{bmatrix} n \\ l \end{bmatrix} - \nu_n([A]_l),$$

so that

$$\frac{\nu_n(A)}{\begin{bmatrix} n \\ k \end{bmatrix}} + 1 - \frac{\nu_n([A]_l)}{\begin{bmatrix} n \\ l \end{bmatrix}} \leq 1.$$

It follows that

$$\nu_n([A]_l) \geq \begin{bmatrix} n \\ l \end{bmatrix} \begin{bmatrix} n \\ k \end{bmatrix}^{-1} \nu_n(A) = \frac{[k]![n-k]!}{[l]![n-l]!} \nu_n(A).$$

$\square$

For example, in the case $l = k - 1$, we have

$$\nu_n([A]_{k-1}) \geq \frac{[k]}{[n-k+1]} \nu_n(A).$$

Theorem 6.4.4 can also be expressed in the language of simplicial complexes. As for the discrete lattice $P(S)$, define $A \subseteq \mathrm{Mod}(n)$ to be a *simplicial complex*, if whenever $V \in A$ and $W \subseteq V$ we have $W \in A$. (Such a set A is also called an *order ideal* of $\mathrm{Mod}(n)$.) In other words, if $V \in A$, then all subspaces of V are also elements of A. Once again the set of maximal elements of a simplicial complex is an antichain, and a simplicial complex having exactly one maximal element is called a *simplex* (or *principal order ideal*). The elements of dimension k in a simplicial complex A are called the k-faces of A. Theorem 6.4.4 can now be thought of as giving a lower bound for the measure of the set of l-faces of A, given the measure of the set of k-faces of A.

6.5 A continuous analogue of Meshalkin's theorem

We now turn to the continuous (vector space) analogues of the r-decompositions and s-systems on the discrete lattice $P(S)$; see also Section 3.1.

A map $\delta : \{1, \ldots, r\} \longrightarrow \mathrm{Mod}(n)$ is called an r-*decomposition of* $\mathbf{R}^n$ if

(i) $\delta(i) \perp \delta(j)$ for $i \neq j$, and
(ii) $\delta(1) \oplus \cdots \oplus \delta(r) = \mathbf{R}^n$

Denote by $\mathrm{Dec}(n, r)$ the set of all r-decompositions of $\mathbf{R}^n$. Note that, for each $\delta \in \mathrm{Dec}(n, r)$,

$$\dim \delta(1) + \cdots + \dim \delta(r) = n.$$

Given positive integers $a_1, a_2, \ldots, a_r$ such that $a_1 + \cdots + a_r = n$, we shall denote by $\mathrm{Mult}(n; a_1, \ldots, a_r)$ the set of all r-decompositions δ such that $\dim \delta(i) = a_i$ for $i = 1, \ldots, r$. In other words, $\mathrm{Mult}(n; a_1, \ldots, a_r)$ is the set of all (ordered) decompositions of $\mathbf{R}^n$ into direct sums of subspaces having dimensions $a_1, \ldots, a_r$. Evidently the set $\mathrm{Dec}(n, r)$ can be expressed as the finite disjoint union

$$\mathrm{Dec}(n, r) = \biguplus_{a_1 + \cdots + a_r = n} \mathrm{Mult}(n; a_1, \ldots, a_r).$$

An s-*system* of order r is a subset $\sigma \subseteq \mathrm{Dec}(n, r)$ such that the set

$$\{\delta(i) : \delta \in \sigma\} \tag{6.13}$$

is an antichain in $\mathrm{Mod}(n)$, for each $1 \leq i \leq r$.

An obvious example of an s-system of order r is $\mathrm{Mult}(n; a_1, \ldots, a_r)$ for some admissible selection of $a_1, \ldots, a_r$. If $\delta, \zeta \in \mathrm{Mult}(n; a_1, \ldots, a_r)$ then $\delta(i)$ and $\zeta(i)$ both have dimension a_i, so that either $\delta(i) = \zeta(i)$ or the two subspaces are incomparable in the subset partial ordering on $\mathrm{Mod}(n)$. This holds for $i = 1, \ldots, r$, and so the antichain condition on (6.13) is satisfied.

Other disguised examples with which we have already worked are the
s-systems of order 2. Let A be an antichain in $\text{Mod}(n)$. For each $V \in A$
we can express $\mathbf{R}^n$ as the direct sum $V \oplus V^\perp$, so that the pair $(V, V^\perp)$
is a 2-decomposition in $\text{Dec}(n, 2)$. Moreover, the set

$$\{V^\perp : V \in A\}$$

is also an antichain in $\text{Mod}(n)$, so that the set

$$\sigma = \{(V, V^\perp) : V \in A\}$$

is an s-system of order 2. Thus the notion of s-system is again a
generalization of the notion of an antichain. Similarly, the Grassmannian $\text{Gr}(n, k)$ can be viewed as $\text{Mult}(n; k, n - k)$ through the bijection
$V \mapsto (V, V^\perp)$.

In analogy to the construction of the measure ν_k^n on $\text{Gr}(n, k)$, define
invariant measures on the sets $\text{Mult}(n; a_1, \ldots, a_r)$ as follows. For $\delta \in$
$\text{Mult}(n; a_1, \ldots, a_r)$ define a flag $(x_0, x_1, \ldots, x_n) \in \text{Flag}(n)$ to be *compatible* with δ if

(i) $x_{a_1} = \delta(1)$, and
(ii) $x_{a_1 + \cdots + a_i}/x_{a_1 + \cdots + a_{i-1}} = \delta(i)$, for $i \geq 2$.

Here the quotient $x_{a_1 + \cdots + a_i}/x_{a_1 + \cdots + a_{i-1}}$ denotes the orthogonal complement of the vector space $x_{a_1 + \cdots + a_{i-1}}$ inside the larger space $x_{a_1 + \cdots + a_i}$.

Let $\text{Flag}(A)$ be the set of all flags $(x_0, x_1, \ldots, x_n)$ compatible with
some $\delta \in A$ for $A \subseteq \text{Mult}(n; a_1, \ldots, a_r)$. Define

$$\nu_{a_1, a_2, \ldots, a_r}^n (A) = \frac{1}{[a_1]! [a_2]! \cdots [a_r]!} \phi_n(\text{Flag}(A)). \tag{6.14}$$

To justify this normalization combinatorially, note that, for each $\delta \in A$,
there are exactly $[a_1]! [a_2]! \cdots [a_r]!$ flags compatible with δ; that is, to
choose a flag compatible with δ one must choose a frame for each of the
vector spaces δ_i, of which there are $[a_i]!$ choices for each i.

The measure $\nu_{a_1, \ldots, a_r}^n$ is evidently invariant under rotations, and we
have

$$\nu_{a_1, \ldots, a_r}^n (\text{Mult}(n; a_1, \ldots, a_r)) = \frac{[n]!}{[a_1]! \cdots [a_r]!} = \begin{bmatrix} n \\ a_1, \ldots, a_r \end{bmatrix}.$$

These values, called *multiflag coefficients*, are continuous analogues of
the multinomial coefficients of the discrete lattice in Chapter 3.

Define a measure $\nu_{n;r}$ on $\text{Dec}(n, r)$ by taking the direct sum of the measures $\nu_{a_1,\ldots,a_r}^n$. That is, for any measurable subset $A \subseteq \mathrm{Dec}(n,r)$, define

$$\nu_{n;r}(A) = \sum_{a_1+\cdots+a_r=n} \nu_{a_1,\ldots,a_r}^n (A \cap \mathrm{Mult}(n;a_1,\ldots,a_r)). \qquad (6.15)$$

Just as the continuous Sperner Theorem 6.4.2 gives the maximum possible measure for an antichain A in $\mathrm{Mod}(n)$, a generalization of this theorem gives the maximum possible measure for an s-system in $\mathrm{Dec}(n,r)$. *En route* to such a generalization we prove a multinomial version of the continuous L.Y.M. inequality. (Compare with Theorem 3.1.4.)

Theorem 6.5.1 (The continuous multinomial L.Y.M. inequality) *Let $\sigma \subseteq \mathrm{Dec}(n,r)$ be an s-system. For $a_1 + \cdots + a_r = n$ denote*

$$\sigma_{a_1,\ldots,a_r} = \sigma \cap \mathrm{Mult}(n;a_1,\ldots,a_r),$$

so that

$$\sigma = \bigcup_{a_1+\cdots+a_r=n} \sigma_{a_1,\ldots,a_r},$$

is a disjoint union. Then

$$\sum_{a_1+\cdots+a_r=n} \frac{\nu_{a_1,\ldots,a_r}^n(\sigma_{a_1,\ldots,a_r})}{\left[\begin{smallmatrix} n \\ a_1,\ldots,a_r \end{smallmatrix}\right]} \le 1. \qquad (6.16)$$

Proof For $a_1+\cdots+a_r = n$ the measure of flags compatible with $\sigma_{a_1,\ldots,a_r}$ is given by

$$\phi_n(\mathrm{Flag}(\sigma_{a_1,\ldots,a_r})) = \nu_{a_1,\ldots,a_r}^n(\sigma_{a_1,\ldots,a_r})[a_1]! \cdots [a_r]!$$

by the definition (6.14) of $\nu_{a_1,\ldots,a_r}^n$.

Suppose that a flag $(x_0,x_1,\ldots,x_n)$ is compatible with both $\gamma,\delta \in \sigma$. Then $\gamma(1) = x_{a_1}$ and $\delta(1) = x_{b_1}$, where $a_1 = \dim\gamma(1)$ and $b_1 = \dim\delta(1)$. Since $(x_0,x_1,\ldots,x_n)$ is a flag, we have $x_{a_1} \subseteq x_{b_1}$ or vice versa. However, σ is an s-system, so that either $\gamma(1) = \delta(1)$ or the two spaces are incomparable. Therefore $\gamma(1) = \delta(1)$ and $a_1 = b_1$. Continuing, we have $\gamma(2) = x_{a_1+a_2}/x_{a_1}$ and $\delta(2) = x_{b_1+b_2}/x_{a_1}$ (since $a_1 = b_1$). A similar argument then implies that $\gamma(2) = \delta(2)$ and $a_2 = b_2$. Continuing in this manner we conclude that $\gamma(i) = \delta(i)$ for each $1 \le i \le r$, so that $\gamma = \delta$. In other words, every flag in $\mathrm{Flag}(n)$ is compatible with at most one r-decomposition $\delta \in \sigma$. It follows that

$$\sum_{a_1+\cdots+a_r=n} \nu_{a_1,\ldots,a_r}^n(\sigma_{a_1,\ldots,a_r})[a_1]! \cdots [a_r]!$$

$$= \sum_{a_1+\cdots+a_r=n} \phi_n(\mathrm{Flag}(\sigma_{a_1,\ldots,a_r})) = \phi_n(\mathrm{Flag}(\sigma))$$

$$\leq [n]!,$$

so that

$$\sum_{a_1+\cdots+a_r=n} \frac{\nu^n_{a_1,\ldots,a_r}(\sigma_{a_1,\ldots,a_r})}{\left[\begin{matrix} n \\ a_1,\ldots,a_r \end{matrix}\right]} \leq 1.$$

□

The multiflag coefficients also satisfy the following property, in analogy to the classical multinomial coefficients.

Proposition 6.5.2 *Let $r \leq n$ be positive integers, and suppose that $n = rq + b$, where $q \geq 0$ is an integer and $0 \leq b \leq r - 1$ is the integer remainder. For $a_1 + \cdots + a_r = n$,*

$$\begin{bmatrix} n \\ a_1,\ldots,a_r \end{bmatrix} \leq \begin{bmatrix} n \\ \underbrace{\langle n/r\rangle,\cdots,\langle n/r\rangle}_{r-b}, \underbrace{\langle n/r\rangle+1,\cdots,\langle n/r\rangle+1}_{b} \end{bmatrix}.$$

Proof Let $a_1,\ldots,a_r$ be positive integers such that $a_1 + \cdots + a_r = n$. Without loss of generality, suppose that $a_1 < \langle n/r \rangle$. Then $a_i > \langle n/r \rangle$ for some $i > 1$. Again without loss of generality, suppose that

$$a_1 < \langle n/r \rangle < a_2.$$

Then $a_2 - a_1 \geq 2$, so that

$$a_1 < \left\langle \frac{a_1 + a_2}{2} \right\rangle < a_2.$$

It then follows from Proposition 6.3.5 that

$$[a_1 + 1]![a_2 - 1]! \leq [a_1]![a_2]!$$

Replace a_1 with $a_1 + 1$ and a_2 with $a_2 - 1$. Note that the identity $a_1 + \cdots + a_r = n$ is preserved. This process is repeated until $a_i \geq \langle n/r \rangle$ for all $1 \leq i \leq r$; that is, until $a_i = \langle n/r \rangle + 1$ for $1 \leq i \leq b$ and $a_i = \langle n/r \rangle$ for $b + 1 \leq i \leq r$, where b is the integer remainder upon division of n by r.

Since each iteration of this procedure decreases the value of the product $[a_1]!\cdots[a_r]!$, it follows that

$$[a_1]!\cdots[a_r]! \geq \left([\langle n/r\rangle]!\right)^{r-b} \left([\langle n/r\rangle + 1]!\right)^{b},$$

for all $a_1 + \cdots + a_r = n$. Therefore,

$$\frac{[n]!}{[a_1]! \cdots [a_r]!} \leq \frac{[n]!}{([\langle n/r \rangle]!)^{r-b} ([\langle n/r \rangle + 1]!)^b},$$

for all $a_1 + \cdots + a_r = n$. □

We are now able to prove a continuous analogue to Meshalkin's Theorem 3.1.5, a multinomial generalization of the continuous Sperner Theorem 6.4.2.

Theorem 6.5.3 (The continuous Meshalkin theorem) *Let σ be an s-system in $\mathrm{Dec}(n, r)$. Then*

$$\nu_{n;r}(\sigma) \leq \left[\underbrace{\langle n/r \rangle, \cdots, \langle n/r \rangle}_{r-b}, \underbrace{\langle n/r \rangle + 1, \cdots, \langle n/r \rangle + 1}_{b} \right],$$

where $n \equiv b \bmod r$.

Proof We reason in analogy to the proof of Meshalkin's Theorem 3.1.5. For $a_1 + \cdots + a_r = n$ let $\sigma_{a_1,\ldots,a_r} = \sigma \cap \mathrm{Mult}(n; a_1, \ldots, a_r)$. By combining (6.16) and Proposition 6.5.2 we obtain

$$\sum_{a_1 + \cdots + a_r = n} \frac{\nu^n_{a_1,\ldots,a_r}(\sigma_{a_1,\ldots,a_r})}{\left[\begin{matrix} n \\ \langle n/r \rangle, \cdots, \langle n/r \rangle, \langle n/r \rangle + 1, \cdots, \langle n/r \rangle + 1 \end{matrix} \right]}$$

$$\leq \sum_{a_1 + \cdots + a_r = n} \frac{\nu^n_{a_1,\ldots,a_r}(\sigma_{a_1,\ldots,a_r})}{\left[\begin{matrix} n \\ a_1,\ldots,a_r \end{matrix} \right]} \leq 1,$$

so that

$$\nu_{n;r}(\sigma) = \sum_{a_1 + \cdots + a_r = n} \nu^n_{a_1,\ldots,a_r}(\sigma_{a_1,\ldots,a_r})$$

$$\leq \left[\begin{matrix} n \\ \langle n/r \rangle, \cdots, \langle n/r \rangle, \langle n/r \rangle + 1, \cdots, \langle n/r \rangle + 1 \end{matrix} \right].$$

□

6.6 Helly's theorem for subspaces

Using the language of simplicial complexes in $\mathrm{Mod}(n)$ introduced in Section 6.4, it may be possible to apply the Euler characteristic techniques of Section 5.3 to prove the following Helly-type theorem for subspaces of $\mathbf{R}^n$. However, this theorem also follows easily from elementary linear algebra.

Theorem 6.6.1 (Helly's theorem for $\mathrm{Mod}(n)$**)** *Let F be a family of nonzero subspaces of $\mathbf{R}^n$. Suppose that, for any subset $G \subseteq F$ such that $|G| \le n$ (that is, every subset of cardinality at most n of F),*

$$\dim \left(\bigcap_{x \in G} x \right) > 0.$$

Then

$$\dim \left(\bigcap_{x \in F} x \right) > 0.$$

In other words, if every n elements of F contain a common line through the origin, then there is at least one line ℓ contained in all of the subspaces in F.

Proof To begin, suppose that F is a *finite* family of subspaces. For this case, the proof is by induction on the size $|F|$ of the family F. If $|F| \le n$ then the theorem holds trivially. Suppose that the theorem holds for the case $|F| = m$ for some $m \ge n$. We then consider the case of $|F| = m+1$.

Write $F = \{x_1, x_2, \ldots, x_{m+1}\}$, and denote

$$y_i = \bigcap_{j \neq i} x_j. \tag{6.17}$$

Since the theorem is true for families of size m, each y_i has positive dimension. That is, for each $i \in \{1, \ldots, m+1\}$ there exists a nonzero vector $v_i \in y_i$. Since $m \ge n$, the collection $\{v_1, v_2, \ldots, v_{m+1}\}$ must be linearly dependent. Without loss of generality, we may then assume that

$$v_{m+1} = c_1 v_1 + \cdots + c_m v_m,$$

where not all the coefficients c_i are zero. However, (6.17) implies that $v_i \in x_{m+1}$ for all $i \in \{1, \ldots, m\}$. It follows that $v_{m+1} \in x_{m+1}$ as well. Since $v_{m+1} \in y_{m+1}$, we have

$$v_{m+1} \in \bigcap_{j=1}^{m+1} x_j.$$

This completes the induction step and the proof of the finite case.

To prove the general case, consider the set $\tilde{x}$ of all lines through the origin contained in a given (nonzero) subspace x; that is

$$\tilde{x} = \{\ell \in \mathrm{Gr}(n, 1) : \ell \subseteq x\}.$$

For all nonzero $x \in \text{Mod}(n)$, the set $\widetilde{x}$ is a closed subset of the compact space $\text{Gr}(n, 1)$. Define

$$\widetilde{F} = \{\widetilde{x} : x \in F\},$$

and define $\widetilde{G}$ similarly for subfamilies G of F.

Suppose G is any *finite* subfamily of subspaces in F. Since F satisfies the intersection condition of Theorem 6.6.1, so does the subfamily G. Since G is finite, it follows from the previous argument that

$$\dim\left(\bigcap_{x \in G} x\right) > 0.$$

In other words, the family $\widetilde{G}$ of closed sets has a non-empty intersection, for every finite subfamily $\widetilde{G}$ of $\widetilde{F}$. It follows from the compactness of $\text{Gr}(n, 1)$ that the family $\widetilde{F}$ of closed sets has a non-empty intersection. Equivalently, the intersection of all subspaces in F has positive dimension. $\square$

6.7 Notes

In this chapter we have assumed the existence and uniqueness of Haar (orthogonal invariant) measures on the spaces $\text{Gr}(n, i)$ and $\text{Flag}(n)$, which are homogeneous spaces of the orthogonal group $O(n)$. For a complete discussion of Haar measures on Lie groups and their homogeneous spaces, see [76]. See also [97, p. 151].

The measures τ_n, ϕ_n, ν_n, and $\nu_{n;r}$ may also be defined from the perspective of the action of $O(n)$ on the domains of these measures. Choosing a normalization for τ_n determines a normalization for the $O(n)$-invariant measure on the group $O(n)$ itself. The actions of $O(n)$ on the spaces $\text{Flag}(n)$, $\text{Mod}(n)$, and $\text{Dec}(n, r)$ then induce precisely the invariant measures ϕ_n, ν_n, and $\nu_{n;r}$ on these spaces. We give a brief sketch of this approach, leaving the details to the reader (see also [76]).

To begin, recall that the columns of an orthogonal $n \times n$ matrix A determine a unique orthogonal frame for $\mathbf{R}^n$. Conversely, given an orthogonal frame in $\mathbf{R}^n$, we can construct an orthogonal matrix by choosing a unit vector from each line in the frame. Since each line in an orthogonal frame contains two unit vectors, there is a 2^n-to-1 correspondence between the group $O(n)$ and the set of frames for $\mathbf{R}^n$, and consequently between $O(n)$ and $\text{Flag}(n)$. Once again setting $[n] = \tau_n(\text{Gr}(n, 1))$, we can construct an $O(n)$-invariant measure $\widetilde{\phi}_n$ on the group $O(n)$ itself, such that

$$\widetilde{\phi}_n(O(n)) = 2[n]\widetilde{\phi}_{n-1}(O(n-1)),$$

so that

$$\widetilde{\phi}_n(O(n)) = 2^n[n]!.$$

Recall that the group $O(n)$ acts transitively on $\mathrm{Gr}(n,k)$, for $0 \leq k \leq n$. Moreover, the stabilizer $\mathrm{Stab}(V)$ of a subspace $V \in \mathrm{Gr}(n,k)$ under this action is isomorphic to the product group $O(k) \times O(n-k)$. This follows from an argument similar to the 'combinatorial' argument on frames given in Section 6.1. It follows from elementary Lie group theory that there exists a diffeomorphism between the space $\mathrm{Gr}(n,k)$ and the quotient *space* (no longer a group)

$$\frac{O(n)}{O(k) \times O(n-k)}.$$

This diffeomorphism induces an invariant measure ν_k^n on $\mathrm{Gr}(n,k)$, such that

$$\nu_k^n(\mathrm{Gr}(n,k)) = \frac{\widetilde{\phi}_n(O(n))}{\widetilde{\phi}_k(O(k))\widetilde{\phi}_{n-k}(O(n-k))} = \frac{[n]!}{[k]![n-k]!},$$

as in our original approach. A similar construction works for the measures $\nu_{n;a_1,\ldots,a_r}$ on the spaces $\mathrm{Mult}(n;a_1,\ldots,a_r)$, and again this construction via group actions agrees with that of Section 6.5.

Theorems 6.4.2, 6.4.3 and 6.5.3, which give continuous analogues of Sperner's and Meshalkin's theorems, are due to Klain and Rota [55].

Many results in the lattice theory of $\mathrm{Mod}(n)$ hold independently of the normalization of the measure τ_n; that is, the value of $[n]$. For example, in [24], Fisk developed a similar construction, in which the total measure of $\mathrm{Gr}(n,1)$ is taken to be $n\omega_n/2$, the measure suggested by the usual two-to-one quotient map from the unit sphere in $\mathbf{R}^n$. However, this normalization does not permit an extension of the lattice analogy sufficient to admit analogues of Sperner's and Meshalkin's theorems, since the value of $n\omega_n/2$ is not an increasing function of n. Moreover, our choice (6.1) of normalization $[n]$ in Section 6.1 is compatible with Cauchy's surface area formula in the sense described by the equation (5.11). In subsequent sections we shall see that (6.1) is in fact the *unique* choice of normalization that agrees with a fundamental normalization for the rigid motion invariant measures on $\mathbf{R}^n$ and its subspaces, namely, the intrinsic volumes (see also [55]).

The bound given in Theorem 6.4.4 is probably not the best. In the discrete case, Katona [48] and Kruskal [57] independently obtained a much stronger result (see Section 3.4), which unlike Theorem 6.4.4 did not depend on the size of the ambient set (or, in the language of subspaces, the dimension n of the ambient vector space). Perhaps a continuous analogue to the Katona–Kruskal theorem is waiting in the wings.

7

The intrinsic volumes for polyconvex sets

In this chapter we shift our attention from the Grassmannian (k-planes passing through the origin in $\mathbf{R}^n$) to the affine Grassmannian (all k-planes in $\mathbf{R}^n$), in order to extend the intrinsic volumes of Section 4.2 from the lattice of parallelotopes to the larger lattice of polyconvex sets. A crucial tool in this generalization will be the Euler characteristic of Section 5.2, which will serve as a device for testing whether a compact convex set intersects a given k-plane. Section 7.4 concludes the chapter with a preliminary version of the mean projection formula, which provides a fundamental connection between the intrinsic volumes and the lattice of subspaces. An improved version of the mean projection formula will appear in Section 9.4.

7.1 The affine Grassmannian

We next consider the partially ordered set $\mathrm{Aff}(n)$ of all linear varieties in $\mathbf{R}^n$, whether through the origin or not. The partially ordered set $\mathrm{Aff}(n)$ is not a well-behaved lattice. The group E_n of Euclidean motions acts naturally on $\mathrm{Aff}(n)$. Denote by $\mathrm{Graff}(n,k)$ the subset of $\mathrm{Aff}(n)$ consisting of all elements of rank k; that is, all linear varieties of dimension k. The minimal element of the partially ordered set $\mathrm{Aff}(n)$ is the empty set $\emptyset$ (unlike in $\mathrm{Mod}(n)$, where the minimal element was the zero subspace $\{0\}$).

We shall prove the existence of a measure λ_k^n on $\mathrm{Graff}(n,k)$ that is invariant under the Euclidean group E_n. To this end, we parametrize $\mathrm{Graff}(n,k)$ as follows. Given $V \in \mathrm{Graff}(n,k)$, let $V^\perp$ be the maximal linear subspace of $\mathbf{R}^n$ orthogonal to V and containing the origin. There is a unique maximal linear subspace $\mathrm{or}(V)$ orthogonal to $V^\perp$ and containing the origin. The subspace $\mathrm{or}(V)$ is of dimension k; we shall say that V and or(V) are *parallel*. The set $V \cap V^\perp$ is a point in V which we denote by $p(V)$.

Thus, to every $V \in \mathrm{Graff}(n,k)$ there corresponds a pair $(\mathrm{or}(V),p)$, where or$(V) \in \mathrm{Gr}(n,k)$ and $p \in \mathrm{or}(V)^\perp \subseteq \mathbf{R}^n$. (Note that $\mathrm{or}(V)^\perp = V^\perp$.) Conversely, given any pair (V_0,p), where $V_0 \in \mathrm{Gr}(n,k)$, and $p \in V_0^\perp$, there is a unique linear variety $V \in \mathrm{Graff}(n,k)$ such that $\mathrm{or}(V) = V_0$ and $V \cap V_0^\perp = p$.

For $V \in \mathrm{Graff}(n,k)$ and $p \in \mathbf{R}^n$, denote by $V + p$ the translation of the linear variety V by the vector p. If f is a real-valued measurable function on $\mathrm{Graff}(n,k)$, let $\bar{f} \colon \mathrm{Gr}(n,k) \times \mathbf{R}^n \longrightarrow \mathbf{R}$ be given by the equation

$$\bar{f}(V_0,p) = f(V_0 + p),$$

and define

$$\int f \, d\lambda_k^n = \int_{\mathrm{Gr}(n,k)} \int_{V_0^\perp} \bar{f}(V_0,p) \, dp \, d\nu_k^n(V_0),$$

where dp denotes the ordinary Lebesgue measure on $V_0^\perp \cong \mathbf{R}^{n-k}$.

We thereby define a measure λ_k^n on $\mathrm{Graff}(n,k)$ that is invariant under the group E_n. To see this, suppose that $g_v \in E_n$ corresponds to translation by the vector v. Then the composition of functions

$$\overline{f \circ g_v}(V_0,p) = f \circ g_v(V_0 + p) = f(V_0 + p + v) = f(V_0 + p + v|_{V_0^\perp}),$$

where $v|_{V_0^\perp}$ denotes the orthogonal projection of the vector v onto the subspace $V_0^\perp$. We then have

$$\begin{aligned}
\int f \circ g_v \, d\lambda_k^n &= \int_{\mathrm{Gr}(n,k)} \int_{V_0^\perp} f(V_0 + p + v|_{V_0^\perp}) \, dp \, d\nu_k^n(V_0) \\
&= \int_{\mathrm{Gr}(n,k)} \int_{V_0^\perp} f(V_0 + p) \, dp \, d\nu_k^n(V_0) \\
&= \int f \, d\lambda_k^n,
\end{aligned}$$

by the translation invariance of the measure dp. A similar argument shows that λ_k^n is invariant under rotations (and reflections), and consequently is rigid motion invariant.

7.2 The intrinsic volumes and Hadwiger's formula

We now consider the relationship between the intrinsic volumes, as defined in Chapter 4, and the invariant measure λ_k^n defined on $\mathrm{Graff}(n,k)$. Once again, fix an orthogonal coordinate system in $\mathbf{R}^n$, and consider the lattice $\mathrm{Par}(n)$ of finite unions of parallelotopes with sides parallel to this fixed coordinate system.

For $A \subset \mathbf{R}^n$, denote by $\mathrm{Graff}(A; k)$ the set of all $V \in \mathrm{Graff}(n, k)$ such that $A \cap V \neq \emptyset$. The relationship between λ_k^n and the intrinsic volumes is given by the following theorem.

Theorem 7.2.1 *For all parallelotopes P in $\mathrm{Par}(n)$,*

$$\mu_{n-k}(P) = C_k^n \lambda_k^n(\mathrm{Graff}(P; k)) \tag{7.1}$$

where the constants C_k^n depend only on n and k.

Note that the equation (7.1) is asserted only for the case in which P is a *parallelotope* and *not* for arbitrary elements of $\mathrm{Par}(n)$ (i.e. not for all finite unions of parallelotopes). Theorem 7.2.1 immediately suggests the following question: what are the values of the constants C_n^k? In Section 9.3 we will prove the amazing fact that $C_n^k = 1$, for all $n \geq 0$ and all $0 \leq k \leq n$. However, much can be accomplished in the meantime. In the course of proving Theorem 7.2.1, we shall need the following lemma.

Lemma 7.2.2 *Let A and B be compact convex sets in $\mathbf{R}^n$ such that $A \cup B$ is convex. Let V be a linear variety of positive dimension such that $V \cap A \neq \emptyset$ and $V \cap B \neq \emptyset$. Then $V \cap A \cap B \neq \emptyset$.*

Proof The set $V \cap (A \cup B) = (V \cap A) \cup (V \cap B)$ is convex, since it is the intersection of two convex sets.

If $V \cap A \cap B$ is empty, then choose points $a \in (V \cap A) - (V \cap B)$ and $b \in (V \cap B) - (V \cap A)$. Let I denote the straight line segment with endpoints at a and b. Clearly $I \subseteq V$. Since $A \cup B$ is convex, we have $I \subseteq A \cup B$, so that $I = (I \cap A) \cup (I \cap B)$. Since I is connected and both $I \cap A$ and $I \cap B$ are closed, it follows that

$$I \cap (A \cap B) = (I \cap A) \cap (I \cap B) \neq \emptyset,$$

contradicting the assumption that $V \cap A \cap B = \emptyset$. $\qquad \square$

Lemma 7.2.2 can be re-stated as follows. If A and B are compact convex sets such that $A \cup B$ is convex, then for every $k > 0$ we have

$$\mathrm{Graff}(A \cap B; k) = \mathrm{Graff}(A; k) \cap \mathrm{Graff}(B; k),$$

so that

$$\lambda_k^n(\mathrm{Graff}(A \cup B; k))$$
$$= \lambda_k^n(\mathrm{Graff}(A; k)) + \lambda_k^n(\mathrm{Graff}(B; k)) - \lambda_k^n(\mathrm{Graff}(A \cap B; k)). \tag{7.2}$$

We now turn to the proof of the theorem.

Proof of Theorem 7.2.1 To begin, define a function η on parallelotopes by

$$\eta(P) = \lambda_k^n(\mathrm{Graff}(P; k)). \tag{7.3}$$

It follows from (7.2) and Theorem 4.1.3 that η has a unique extension to a valuation on all of $\mathrm{Par}(n)$ (although the equation (7.3) will not hold for arbitrary finite unions of parallelotopes). Since λ_k^n is invariant, so is the valuation η.

Now suppose that P is a parallelotope, and let f_P be the indicator function of $\mathrm{Graff}(P; k)$ in $\mathrm{Graff}(n, k)$. In the language of the preceding section we have

$$\lambda_k^n(\mathrm{Graff}(P; k)) = \int f_P \, d\lambda_k^n = \int \int \bar{f}_P(V_0, p) \, dp \, d\nu_k^n(V_0), \tag{7.4}$$

where p ranges over $V_0^\perp$.

For a fixed $V_0 \in \mathrm{Gr}(n, k)$, we have

$$\bar{f}_P(V_0, p) = f_P(V_0 + p) = 1$$

if and only if $(V_0 + p) \cap P \neq \emptyset$ (otherwise the function takes the value zero); that is, if and only if $p \in P|V_0^\perp$, where $P|V_0^\perp$ denotes the orthogonal projection of P onto the subspace $V_0^\perp$. In other words, the function $\bar{f}_P(V_0, p)$ is the indicator function of $P|V_0^\perp$. For $\alpha > 0$ we now have

$$\begin{aligned}
\lambda_k^n(\mathrm{Graff}(\alpha P; k)) &= \int_{\mathrm{Gr}(n,k)} \int_{V_0^\perp} \bar{f}_{\alpha P}(V_0, p) \, dp \, d\nu_k^n(V_0) \\
&= \int_{\mathrm{Gr}(n,k)} \int_{V_0^\perp} I_{\alpha P|V_0^\perp} \, dp \, d\nu_k^n(V_0),
\end{aligned}$$

so that

$$\lambda_k^n(\mathrm{Graff}(\alpha P; k)) = \int_{\mathrm{Gr}(n,k)} \mathrm{vol}_{n-k}(\alpha P|V_0^\perp) \, d\nu_k^n(V_0), \tag{7.5}$$

where vol_{n-k} denotes the $(n-k)$-dimensional volume in the $(n-k)$-dimensional space $V_0^\perp$. Since $(n-k)$-dimensional volume is homogeneous of degree $n-k$, we can continue:

$$\begin{aligned}
\eta(\alpha P) &= \lambda_k^n(\mathrm{Graff}(\alpha P; k)) \\
&= \alpha^{n-k} \int_{\mathrm{Gr}(n,k)} \mathrm{vol}_{n-k}(P|V_0^\perp) \, d\nu_k^n(V_0) \\
&= \alpha^{n-k} \int_{\mathrm{Gr}(n,k)} \int_{V_0^\perp} I_{P|V_0^\perp} \, dp \, d\nu_k^n(V_0)
\end{aligned}$$

$$= \alpha^{n-k} \lambda_k^n (\text{Graff}(P; k))$$
$$= \alpha^{n-k} \eta(P).$$

It follows that η is homogeneous of degree $n - k$ on parallelotopes, and consequently on all of $\text{Par}(n)$. Since the $(n - k)$-volume is continuous on compact convex sets in $V_0^\perp$, it also follows from (7.5) that η is a continuous valuation. Therefore, by Corollary 4.2.6, there exists a constant $\gamma_k^n \in \mathbf{R}$ such that

$$\eta(P) = \gamma_k^n \mu_{n-k}(P)$$

for all $P \in \text{Par}(n)$. It is clear from (7.5) that $\eta(P) > 0$ if P has a non-empty interior in $\mathbf{R}^n$. It follows that $\gamma_k^n \neq 0$. Setting $C_k^n = 1/\gamma_k^n$, we have

$$\mu_{n-k}(P) = C_k^n \lambda_k^n (\text{Graff}(P; k))$$

for all parallelotopes $P \in \text{Par}(n)$. □

Theorem 7.2.1 relates the intrinsic volumes, as defined in Chapter 4 in terms of product valuations, to the invariant measure λ_k^n on $\text{Graff}(n, k)$. So far our main result asserts that the intrinsic volume μ_{n-k}, which for parallelotopes is evaluated as a symmetric function of the side lengths, also evaluates the 'measure' of the set of k-dimensional planes meeting a given parallelotope. This interpretation extends to convex sets. The obvious idea for the further extension of μ_{n-k} to all polyconvex sets is to use Groemer's extension Theorem 5.1.1; that is, to define a continuous set function μ_{n-k}^n on a compact convex set K by

$$\mu_{n-k}^n(K) = C_k^n \lambda_k^n (\text{Graff}(K; k)). \qquad (7.6)$$

It then follows from (7.2) and (7.5) that μ_{n-k}^n is a continuous valuation on compact convex sets. Theorem 5.1.1 then asserts that μ_{n-k}^n has a unique extension to all polyconvex sets in $\mathbf{R}^n$. Note, however, that the equation (7.6) is valid only if K is *convex*.

Alternatively, we can give a more constructive definition for μ_{n-k}^n on $\text{Polycon}(n)$. It turns out that we can let the Euler characteristic do most of the work. To motivate this method, recall the definition of μ_{n-k} on $\text{Par}(n)$ as a product valuation. Let f be any simple function on $\text{Par}(n)$. We have shown that

$$\int f \, d\mu_{n-k} = \sum_\sigma \int \int \cdots \int f(x_1, x_2, \ldots, x_n) \, d\mu_0^1(x_{\sigma(1)}) \, d\mu_0^1(x_{\sigma(2)})$$

$$\cdots d\mu_0^1(x_{\sigma(k)}) \, d\mu_1^1(x_{\sigma(k+1)}) \cdots d\mu_1^1(x_{\sigma(n)}),$$

where the sum ranges over all permutations σ of the set $\{1, 2, \ldots, n\}$. The integral

$$\int \int \cdots \int f(x_1, x_2, \ldots, x_n) \, d\mu_0^1(x_1) \cdots d\mu_0^1(x_k) \tag{7.7}$$

can be rewritten as follows. Let V be the linear variety of dimension k parallel to the linear subspace spanned by $x_1, x_2, \ldots, x_k$, and passing through the point $0, \ldots, 0, x_{k+1}, \ldots, x_n$. Let $A \in \mathrm{Polycon}(n)$, and let $f = I_A$. Then the integral (7.7) simplifies to

$$\int \int \cdots \int I_A I_V \, d\mu_0^1(x_1) \cdots d\mu_0^1(x_k) = \mu_0(A \cap V).$$

To indicate the dependence of V on the coordinates $x_{k+1}, x_{k+2}, \ldots, x_n$, let us write $V = V(x_{k+1}, \ldots, x_n)$. We then have

$$\int I_A \, d\mu_{n-k} = \sum_\sigma \int \int \cdots \int \mu_0 \left(A \cap V(x_{\sigma(k+1)}, \ldots, x_{\sigma(n)}) \right)$$

$$d\mu_1^1(x_{\sigma(k+1)}) \cdots d\mu_1^1(x_{\sigma(n)}).$$

This formula suggests replacing the summation over the symmetric group of permutations by integration over the space $\mathrm{Gr}(n, k)$ relative to the invariant measure ν_k^n. Specifically, let $V(V_0; p)$ be the linear variety of dimension k parallel to the linear subspace V_0, and intersecting the linear subspace $V_0^\perp$ at the point $p \in \mathbf{R}^n$. We then obtain

$$\int \int \mu_0 \left(A \cap V(V_0; p) \right) d\mu_{n-k}^{n-k}(p) \, d\nu_k^n(V_0).$$

This can be simplified further, since the integration is practically carried out over the invariant measure on $\mathrm{Graff}(n, k)$. For all $A \in \mathrm{Polycon}(n)$, define

$$\widetilde{\mu}_{n-k}^n(A) = C_k^n \int \mu_0(A \cap V) \, d\lambda_k^n(V). \tag{7.8}$$

This formula is known as *Hadwiger's formula*. The integral ranges over all $V \in \mathrm{Graff}(n, k)$. It is easily verified that μ_{n-k}^n, as extended by (7.8), is a convex-continuous invariant valuation on $\mathrm{Polycon}(n)$.

Note that $\mu_0(K \cap V) = I_{\mathrm{Graff}(K;k)}(V)$ if K is a compact convex set. Hadwiger's formula (7.8) then implies that

$$\widetilde{\mu}_{n-k}^n(K) = C_k^n \lambda_k^n(\mathrm{Graff}(K; k)).$$

Thus, the valuation $\widetilde{\mu}_{n-k}^n$ defined by (7.8) agrees with the alternative extension (7.6) as well as with Theorem 7.2.1. However, the equation (7.8) defines intrinsic volumes on every polyconvex set A, giving the desired extension explicitly.

Recall that the intrinsic volume μ_k was called 'intrinsic,' because $\mu_k(P)$ remains the same for a parallelotope P, even if P is rigidly embedded within a higher dimensional space. In other words, the valuations μ_k on $\mathrm{Par}(n)$ are normalized independently of n (see Theorem 4.2.2). We shall see later that this property carries over to the extension of μ_k given by (7.8), so that our use of the term 'intrinsic' remains justified (see Theorem 8.4.1). One way to see this is to let $K \in \mathcal{K}^n$, embed $\mathbf{R}^n$ into $\mathbf{R}^{n+1}$, and compare $\mu_k^n(K)$ with $\mu_k^{n+1}(K)$ as given by (7.8). The value of μ_k^{n+1} is given as the measure of all $(n-k+1)$-dimensional planes in $\mathbf{R}^{n+1}$ that meet K. Since a generic $(n-k+1)$-plane in $\mathbf{R}^{n+1}$ meets $\mathbf{R}^n$ in a plane of dimension $n-k$, the measure of all $(n-k+1)$-planes meeting K in $\mathbf{R}^{n+1}$ is proportional to the number of $(n-k)$-planes meeting K in $\mathbf{R}^n$, where the proportionality constant is independent of K, and is given by the measure (in the appropriate sense) of $(n-k+1)$-planes in $\mathbf{R}^{n+1}$ containing a given (fixed) $(n-k)$-plane. If we call this proportionality constant $\alpha_k^{n,n+1}$, we have

$$\mu_k^{n+1}(K) = \alpha_k^{n,n+1}\mu_k^n(K)$$

for all $K \in \mathrm{Polycon}(n)$. Meanwhile, Theorem 4.2.2 asserts $\mu_k^{n+1}(P) = \mu_k^n(P)$ for all $P \in \mathrm{Par}(n)$. It follows that $\alpha_k^{n,n+1} = 1$, and the intrinsic volumes are universally normalized.

The problem in this argument is the claim that the measure of all $(n-k+1)$-planes meeting K in $\mathbf{R}^{n+1}$ is proportional to the number of $(n-k)$-planes meeting K in $\mathbf{R}^n$. This statement, although true, requires additional measure-theoretic detail in order to be made precise. Instead, we defer a rigorous proof of the universal normalization to Section 8.4.

For the interim, and in order to avoid ambiguity, we will continue to use the notation μ_k^n for the kth intrinsic volume on $\mathrm{Polycon}(n)$ until we are able to prove the normalization property rigorously in Theorem 8.4.1. We will make an exception for μ_0, which is clearly normalized independently of the ambient space $\mathbf{R}^n$, since $\mu_0(K) = 1$ for all non-empty compact convex sets K.

We are now in a position to generalize the results of Section 1.2. Let K and L be compact convex sets in $\mathbf{R}^n$. Suppose that L has dimension n, and that $K \subseteq L$. Then $\mathrm{Graff}(K;k) \subseteq \mathrm{Graff}(L;k)$, and the conditional probability that a linear variety of dimension k shall meet K, given that it meets L, is given by the ratio

$$\frac{\lambda_k^n(\mathrm{Graff}(K;k))}{\lambda_k^n(\mathrm{Graff}(L;k))}.$$

From (7.6) and (7.8) we now deduce the following.

Theorem 7.2.3 (Sylvester's theorem) *Let $K \subseteq L$ be compact convex sets. Suppose that L is of dimension n. The conditional probability that a linear variety of dimension k shall meet K, given that it meets L, is given by*

$$\frac{\mu_{n-k}^n(K)}{\mu_{n-k}^n(L)}.$$

□

Note that this probability can be computed relative to any invariant measure on $\mathrm{Graff}(n,k)$. The normalizing constants C_k^n are irrelevant, since they cancel in the ratio above.

7.3 An Euler relation for the intrinsic volumes

The Euler characteristic formula of Theorem 5.2.4 generalizes to intrinsic volumes. To this end, we recall from Theorem 5.2.3 that the Euler characteristic for the interior of a convex polytope P *of dimension n* is given by

$$\mu_0(\mathrm{int}\ P) = (-1)^n.$$

Let P be a compact convex polytope. For almost all $V \in \mathrm{Graff}(n, n-k)$ (with respect to the measure λ_{n-k}^n), we have $\mathrm{relint}(P) \cap V \neq \emptyset$ whenever $P \cap V \neq \emptyset$. Hadwiger's formula (7.8) then implies that

$$
\begin{aligned}
&\mu_k^n(\mathrm{relint}(P)) \\
&= C_k^n \int_{V \in \mathrm{Graff}(n,n-k)} \mu_0(\mathrm{relint}(P) \cap V)\, \mathrm{d}\lambda_{n-k}^n(V) \\
&= C_k^n \int_{V \in \mathrm{Graff}(n,n-k)} (-1)^{\dim(\mathrm{relint}(P) \cap V)} \mu_0(P \cap V)\, \mathrm{d}\lambda_{n-k}^n(V).
\end{aligned}
$$

Note that if $\mathrm{relint}(P) \cap V \neq \emptyset$, then

$$\dim(\mathrm{relint}(P) \cap V) = \dim(P \cap V) = \dim P + \dim V - \dim(P \cup V),$$

where $\dim(P \cup V)$ denotes the dimension of the smallest plane containing

$P \cup V$. If $\mu_k^n(\text{relint}(P)) \neq 0$, then $\dim P \geq k$. Therefore, $\dim(P \cup V) = n$ for almost all $V \in \text{Graff}(n, n - k)$. It follows that

$$\dim(\text{relint}(P) \cap V) = \dim P + (n - k) - n = \dim P - k,$$

and we obtain the formula

$$\mu_k^n(\text{relint}(P)) = (-1)^{\dim P - k} \mu_k^n(P) \qquad (7.9)$$

for all convex polytopes P.

From Hadwiger's formula (7.8) we now derive Euler relations for the intrinsic volumes of an arbitrary polytope P. As in Section 5.2, we define a *system of faces* of P to be a family F of compact convex polytopes such that if $Q, Q' \in F$ and $Q \neq Q'$ then $(\text{relint}(Q)) \cap (\text{relint}(Q')) = \emptyset$, and such that

$$\bigcup_{Q \in F} (\text{relint}(Q)) = P.$$

Under these conditions, the formula (7.9) yields at once

$$\mu_k^n(P) = \sum_{Q \in F} (-1)^{\dim Q - k} \mu_k^n(Q).$$

7.4 The mean projection formula

We conclude this section with an alternative interpretation of Hadwiger's formula (7.8) for compact *convex* sets.

Theorem 7.4.1 (The mean projection formula) *For all $K \in \mathcal{K}^n$,*

$$\mu_k^n(K) = C_{n-k}^n \int_{\text{Gr}(n,k)} \mu_k^k(K | V_0) \, d\nu_k^n(V_0).$$

Recall that μ_k^k denotes the k-dimensional volume on each k-dimensional subspace of $\mathbf{R}^n$. Theorem 7.4.1 states that the kth intrinsic volume $\mu_k^n(K)$ is proportional to the mean of the k-volumes of the orthogonal projections of K onto all k-dimensional subspaces of $\mathbf{R}^n$.

Proof For $K \in \mathcal{K}^n$, we have

$$\begin{aligned}
\mu_k^n(K) &= C_{n-k}^n \lambda_{n-k}^n(\text{Graff}(K; n - k)) \\
&= C_{n-k}^n \int_{\text{Gr}(n,n-k)} \int_{V_0^\perp} \bar{f}_K(V_0, p) \, dp \, d\nu_{n-k}^n(V_0)
\end{aligned}$$

Recall that $\bar{f}_K(V_0, p) = 1$ if $K \cap (V_0 + p) \neq \emptyset$ and is zero otherwise. Since $K \cap (V_0 + p) \neq \emptyset$ if and only if p lies inside the projection $K|V_0^\perp$, we have $\bar{f}_K(V_0, p) = I_{K|V_0^\perp}(p)$. Therefore,

$$
\begin{aligned}
\mu_k^n(K) &= C_{n-k}^n \lambda_{n-k}^n(\text{Graff}(K; n-k)) \\
&= C_{n-k}^n \int_{\text{Gr}(n,n-k)} \int_{V_0^\perp} I_{K|V_0}(p) \, dp \, d\nu_{n-k}^n(V_0) \\
&= C_{n-k}^n \int_{\text{Gr}(n,n-k)} \mu_k^k(K|V_0^\perp) \, d\nu_{n-k}^n(V_0) \\
&= C_{n-k}^n \int_{\text{Gr}(n,k)} \mu_k^k(K|V_0) \, d\nu_k^n(V_0),
\end{aligned}
$$

where the last equality follows from the fact that the orthogonal duality between $\text{Gr}(n,k)$ and $\text{Gr}(n, n-k)$ preserves measure.   □

Note that the preceding argument also yields

$$
\lambda_{n-k}^n(\text{Graff}(K; n-k)) = \int_{\text{Gr}(n,k)} \mu_k^k(K|V_0) \, d\nu_k^n(V_0)
$$

for all *polyconvex* sets K. However, since (7.6) defines μ_k^n for *convex* sets only, Theorem 7.4.1 need not hold for all $K \in \text{Polycon}(n)$.

7.5 Notes

A brief survey of Euler relations for valuations on $\text{Polycon}(n)$ appeared in [72, p. 218–220]. See also [37, 39, 79, 80].

For a general reference on the theory of valuations on convex bodies, see [71, 72] and [85]. For additional related integral geometric formulas and applications, see also [82] and [99].

The theory of intrinsic volumes can be extended to partially ordered sets. Let P be a finite partially ordered set, with a unique minimal element o. An *order ideal* of P is a subset A of P such that, if $x \in A$ and $y \le x$, then $y \in A$. The set of all order ideals of P is denoted by $L(P)$. It is closed under unions and intersections; that is, it is a distributive lattice. (It can be shown that every finite distributive lattice can be represented in the form $L(P)$ for some partially ordered set P). We shall consider valuations on $L(P)$. If $x \in P$, denote by $\hat{x}$ the order ideal consisting of all $y \in P$ such that $y \le x$. A valuation μ on P will be called *invariant* if $\mu(A) = \mu(B)$ whenever the order ideals A and B are isomorphic as partially ordered sets.

It can be shown that every valuation μ on $L(P)$ extends uniquely to the Boolean algebra of all subsets of P, and that a valuation is uniquely determined by assigning the value $\mu(\hat{x})$ for each $x \in P$. We typically consider only valuations such that $\mu(\hat{o}) = 0$.

There are always at least two invariant valuations on P, namely, the *size*;

$$\sigma(A) = |A| - 1, \ A \neq \{\hat{o}\},$$

and the *Euler characteristic*, which is uniquely determined by setting

$$\mu_0(\hat{x}) = 1 \text{ if } \hat{x} \neq \{\hat{o}\}.$$

It can be shown (see [78, 79, 80]) that the Euler characteristic is closely related to the *Möbius function* of P; that is, the integer-valued function on P uniquely defined by the conditions

$$\mu(\hat{o}, \hat{o}) = 1$$

$$\sum_{q \leq p} \mu(\hat{o}, q) = 0, \ p > \hat{o}.$$

One finds that

$$\mu_0(A) = - \sum_{p \in A, \ p \neq \hat{o}} \mu(\hat{o}, p).$$

Thus, one obtains a generalization of the expression for the Euler characteristic in terms of the 'number of faces'. Several special cases have already been studied in detail.

Other invariant valuations are defined as follows. Let α be an isomorphism class of 'simplices' $\hat{x}$. For every $A \in L(P)$, let μ_α be the number of $\hat{x}$ of class α contained in A. We conjecture that the μ_α and μ_0 span the vector space of all invariant valuations. The identities holding among the μ_α are particularly interesting to determine.

Actually, the theory of invariant valuations on partially ordered sets should proceed along general lines. There is a more general notion of invariant measure that deals with arbitrary segments $[x, y]$ of P, rather than just segments $\hat{x} = [0, x]$. Such valuations can be multiplied in a way that resembles multiplication in the incidence algebra of P (see [78, 79, 92]).

There is more than one gap in the analogy between the discrete and the continuous case. On the one hand, the ease with which the theory of invariant valuations can be carried over to arbitrary partially ordered sets suggests the possibility of generalization of the intrinsic volumes to a more general space than $\mathbf{R}^n$. On the other hand, the analogy between $L(S)$ (or $P(S)$) and $\mathrm{Mod}(n)$ is deficient, since we do not consider order ideals in Mod(n) (or Aff(n)) as our basic building blocks, but rather the polyconvex sets. The difficulty here is that of singling out a sufficiently ample class of order ideals in Mod(n) that can be taken as simplices (so as to replace convex sets). Convexity on Grassmannians is at present too unwieldy a notion, and somehow the Schubert cell structure of the Grassmannian must be brought to bear on the problem (see, for example, [74]).

8

A characterization theorem for volume

In this chapter we state and prove a fundamental theorem of geometric probability, namely the characterization of volume on polyconvex sets as a continuous rigid motion invariant simple valuation Polycon(n). The characterization of volume will lead in turn to a straightforward characterization for all of the intrinsic volumes (see Section 9.1). In Section 8.4 we verify the universal normalization of the intrinsic volumes. In Section 8.5 we investigate a connection between volume and random motions, leading to an analogue of the Buffon needle problem, in which evenly spaced lines are replaced by a discrete additive subgroup of $\mathbf{R}^n$. The results of Section 8.5 will be used in Section 9.6 to solve the Buffon needle problem in a still more general form.

8.1 Simple valuations on polyconvex sets

In this section we state and prove a characterization theorem for volume, leading to a generalization of Theorem 4.2.4 to Polycon(n). Recall that Theorem 4.2.4 characterized volume in two ways, one involving continuity and the other involving monotonicity. It turns out to be much easier to generalize Theorem 4.2.4 in the monotonic case than it is in the continuous case.

Theorem 8.1.1 *Suppose that μ is a monotone translation invariant simple valuation on* Polycon(n). *Then there exists $c \in \mathbf{R}$ such that $\mu(K) = c\mu_n(K)$, for all $K \in$ Polycon(n).*

Proof We shall assume without loss of generality that μ is an increasing valuation, for, if μ is decreasing, then the valuation $-\mu$ is increasing.

Consider first the restriction of μ to Par(n). By Theorem 4.2.4, there exists $c \in \mathbf{R}$ such that $\mu(P) = c\mu_n(P)$, for all $P \in$ Par(n).

Next, consider $K \in \mathcal{K}^n$. Recall from the definition of volume in elementary calculus that $\mu_n(K)$ is given by the supremum of $\mu_n(P)$ over all $P \in$ Par(n) such that $P \subseteq K$. Since μ is increasing, we have $\mu(P) \leq \mu(K)$ for all $P \subseteq K$. In other words, $c\mu_n(P) \leq \mu(K)$ for all $P \subseteq K$. It follows that $c\mu_n(K) \leq \mu(K)$.

Recall again that $\mu_n(K)$ is also given by the infimum of $\mu_n(P)$ over all $P \in$ Par(n) such that $K \subseteq P$. It then similarly follows that $\mu(K) \leq c\mu_n(K)$. Hence, $\mu(K) = c\mu_n(K)$, for all $K \in \mathcal{K}^n$, and therefore also for all $K \in$ Polycon(n). □

A characterization for volume as a continuous valuation, free of monotonicity conditions, requires considerably more work. We begin with some preliminary definitions.

Recall from Section 5.1 that a non-empty compact convex set $K \in \mathcal{K}^n$ is determined uniquely by its *support function* $h_K : \mathbf{S}^{n-1} \longrightarrow \mathbf{R}$, defined by $h_K(u) = \max_{x \in K}\{x \cdot u\}$, where $\cdot$ denotes the standard inner product on $\mathbf{R}^n$. Recall also that, if $v \in \mathbf{R}^n$ and $\bar{v}$ denotes the line segment with endpoints v and $-v$, then $h_{\bar{v}}(u) = |u \cdot v|$, for all $u \in \mathbf{S}^{n-1}$.

For $K \in \mathcal{K}^n$, denote by $-K$ the set $\{x : -x \in K\}$; that is, the reflection of K through the origin. If $K = -K$ we say that K is *centered* or *symmetric about the origin*. A set K is *centered* or *symmetric* if some translate of K is centered about the origin. Denote by $\mathcal{K}_c^n$ the set of all centered compact convex sets in $\mathbf{R}^n$.

Recall that for compact convex sets K and L the *Minkowski sum* $K+L$ is defined by

$$K + L = \{x + y : x \in K \text{ and } y \in L\},$$

and that $h_{K+L} = h_K + h_L$.

A *zonotope* is a finite Minkowski sum of straight line segments. A convex body Y is called a *zonoid* if Y can be approximated in $\mathcal{K}^n$ by a convergent sequence of zonotopes [85, p. 183]. We shall need the following useful fact concerning zonoids and smooth convex bodies. A complete discussion of this result and its proof may be found in [26, 34, 85].

Proposition 8.1.2 *Let* $K \in \mathcal{K}_c^n$, *and suppose that the support function* h_K *is* C^∞. *Then there exist zonoids* Y_1, Y_2 *such that*

$$K + Y_2 = Y_1.$$

Proof For $g \in C^\infty(\mathbf{S}^{n-1})$, the *cosine transform* of g, denoted Cg, is given by the equation

$$Cg(u) = \int_{\mathbf{S}^{n-1}} |u \cdot v| g(v) \, \mathrm{d}v.$$

The transform C is a bijective linear operator on the space of all even C^∞ functions on $\mathbf{S}^{n-1}$. This fact is a consequence of Schur's lemma (or the Funke–Hecke theorem) for spherical harmonics [85, pp. 182–189] (see also [26, 34]).

Since the function $h_K : \mathbf{S}^{n-1} \longrightarrow \mathbf{R}$ is C^∞ and even, there exists an even C^∞ function $g : \mathbf{S}^{n-1} \longrightarrow \mathbf{R}$ such that $h_K = Cg$; that is,

$$h_K(u) = \int_{\mathbf{S}^{n-1}} |u \cdot v| g(v) \, \mathrm{d}v.$$

Let $g^+(v) = \max\{g(v), 0\}$, and let $g^-(v) = \max\{-g(v), 0\}$. Then

$$h_K(u) + \int_{\mathbf{S}^{n-1}} |u \cdot v| g^-(v) \, \mathrm{d}v = \int_{\mathbf{S}^{n-1}} |u \cdot v| g^+(v) \, \mathrm{d}v. \tag{8.1}$$

It is easy to check that the functions $h_{Y_1} = Cg^+$ and $h_{Y_2} = Cg^-$ each satisfy the properties of a support function of a centered convex body, which we denote by Y_1 and Y_2 respectively. The equation (8.1) is then equivalent to the statement that $K + Y_2 = Y_1$. Moreover, since the Riemann sums converging to the integrals in (8.1) are linear combinations of support functions of line segments (i.e. support functions of zonotopes), it follows that Y_1 and Y_2 are zonoids. $\qquad\square$

Let $SO(n)$ denote the special orthogonal group; that is, the set of all rotations of $\mathbf{R}^n$. Let $\mathcal{B} = \{e_1, \ldots, e_n\}$ denote the standard basis for $\mathbf{R}^n$, and denote by $SO(n, \mathcal{B})$ the set of all rotations in $SO(n)$ that fix at least $n - 2$ elements of the basis $\mathcal{B}$.

Proposition 8.1.3 *Suppose that $\phi \in SO(n)$. Then there exists a finite collection $\phi_1, \phi_2, \ldots, \phi_m \in SO(n, \mathcal{B})$ such that $\phi = \phi_1 \phi_2 \cdots \phi_m$.*

Proof The proposition holds trivially in dimension $n = 2$, since $SO(2, \mathcal{B}) = SO(2)$. Suppose that $n \geq 3$ and that the proposition holds for dimension $n - 1$.

Let $\phi \in SO(n)$, and suppose that $\phi \notin SO(n, \mathcal{B})$. Let $v = \phi e_n$, and assume without loss of generality that $v \neq e_n$. Let v' denote the unit normalization of the orthogonal projection of v onto $\mathrm{Span}\{e_1, \ldots, e_{n-1}\} = \mathbf{R}^{n-1}$. There exists $\psi \in SO(n)$ such that $\psi e_n = e_n$ and $\psi v' = e_{n-1}$. As v lies within $\mathrm{Span}\{v', e_n\}$, it follows that ψv lies within $\mathrm{Span}\{e_{n-1}, e_n\}$.

Let ζ be the rotation that fixes $e_1, \ldots, e_{n-2}$ and rotates ψv to e_n. Then $\zeta \in SO(n, \mathcal{B})$, and $\zeta \psi \phi e_n = \zeta \psi v = e_n$. Let $\eta = \zeta \psi \phi$.

Since ψ and η both fix e_n, it follows from the induction assumption on $SO(n-1)$ that there exist $\psi_1, \ldots \psi_i, \eta_1, \ldots \eta_j \in SO(n, \mathcal{B})$ such that $\psi = \psi_1 \cdots \psi_i$ and $\eta = \eta_1 \cdots \eta_j$. Thus,

$$\phi = \psi^{-1} \zeta^{-1} \eta = \psi_i^{-1} \cdots \psi_1^{-1} \zeta^{-1} \eta_1 \cdots \eta_j.$$

$\square$

A valuation μ on $\mathcal{K}^n$ is said to be *simple* if μ vanishes on sets of dimension less than n.

Theorem 8.1.4 *Suppose that μ is a continuous translation invariant simple valuation on $\mathcal{K}^n$. Suppose also that $\mu([0,1]^n) = 0$, and that $\mu(K) = \mu(-K)$, for all $K \in \mathcal{K}^n$. Then $\mu(K) = 0$, for all $K \in \mathcal{K}^n$.*

Here $[0,1]^n$ denotes the n-fold Cartesian product of the closed unit interval $[0,1]$ with itself; that is, a unit n-cube.

Proof If $n = 1$ then the result follows readily, since a compact convex subset of $\mathbf{R}$ is merely a closed line segment. Since μ is simple and vanishes on the closed line segment $[0,1]$, it must vanish on all closed line segments of rational length. It then follows from continuity that μ vanishes on all closed line segments.

For $n > 1$, assume that Theorem 8.1.4 holds for valuations on $\mathcal{K}^{n-1}$. Since μ is translation invariant and simple, the fact that $\mu([0,1]^n) = 0$ implies that $\mu([0, 1/k]^n) = 0$ for all integers $k > 0$. Therefore, $\mu(C) = 0$ for every box C of rational dimensions, with sides parallel to the coordinate axes. This follows from the fact that such a box can be built up out of cubes of the form $[0, 1/k]^n$ for some $k > 0$. The continuity of μ then implies that $\mu(C) = 0$ for every box C of positive real dimensions, with sides parallel to the coordinate axes.

Next, suppose that D is a box with sides parallel to a different set of orthogonal axes. If $n = 2$ then it is easy to see that D can be cut into a finite number of pieces, *translations* of which can be pasted to form a box C with sides parallel to the original coordinate axes (see Figure 8.1). Since μ is simple and translation invariant, it follows that $\mu(D) = \mu(C) = 0$. If $n > 2$, then for all rotations $\zeta \in SO(n, \mathcal{B})$, a box with sides parallel to the basis $\zeta \mathcal{B}$ can be cut, translated, and re-pasted into a box parallel to $\mathcal{B}$, using precisely the operations followed in the case $n = 2$. This works because the rotation ζ fixes at least $n - 2$ of the original coordinate axes. More generally, for $\psi \in SO(n)$, Proposition 8.1.3 states that ψ is a finite product of elements of $SO(n, \mathcal{B})$.

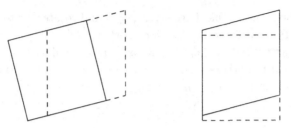

Fig. 8.1. Re-orient a frame without use of rotations.

Therefore, a box with sides parallel to the basis $\psi\mathcal{B}$ can be cut, translated, and re-pasted into a box parallel to $\mathcal{B}$, using a finite iteration of operations of the type used in the case $n = 2$.

It follows that, if D is a box with sides parallel to any orthogonal frame in $\mathbf{R}^n$, then D can be transformed into a box C with sides parallel to the original coordinate axes, by means of cutting, pasting, and translations. Therefore, we have $\mu(D) = \mu(C) = 0$.

Next, define a valuation τ on $\mathcal{K}^{n-1}$ as follows. Given a compact convex subset K of $\mathbf{R}^{n-1}$, set

$$\tau(K) = \mu(K \times [0,1]).$$

Note that $\tau([0,1]^{n-1}) = \mu([0,1]^n) = 0$. Notice also that τ satisfies the hypotheses of Theorem 8.1.4 in dimension $n - 1$. The induction hypothesis then implies that $\tau = 0$.

Since μ is simple, it follows that $\mu(K \times [a,b]) = 0$, for any convex body $K \subseteq \mathbf{R}^{n-1}$ and any rational numbers a and b, with $a \leq b$. The continuity of μ then implies that $\mu(K \times [a,b]) = 0$ for all $a, b \in \mathbf{R}$. Said differently, μ is zero on any *right cylinder* with a convex base.

Let $x_1, \ldots, x_n$ be the coordinates on $\mathbf{R}^n$. We can represent $\mathbf{R}^{n-1}$ by the hyperplane $x_n = 0$. The right cylinders for which we have shown μ to be zero have tops and bottoms that are congruent and that lie directly above and below each other. In other words, the edges connecting the top face to the bottom face are orthogonal to the hyperplane $x_n = 0$.

This process can be applied to right cylinders with base in any $(n-1)$-dimensional subspace of $\mathbf{R}^n$. Since $\mu = 0$ on boxes in every orientation, it follows (from the preceding argument) that $\mu = 0$ on right cylinders of every orientation.

Suppose that M is a *prism*, or slanting cylinder, for which the top and bottom faces are congruent and both parallel to the hyperplane $x_n =$

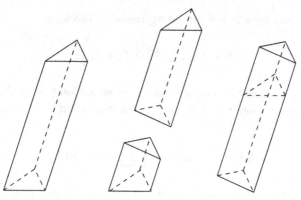

Fig. 8.2. Turn a prism into a right cylinder.

0, but whose cylindrical boundary is no longer orthogonal to $x_n = 0$, meeting it instead at some constant angle. See Figure 8.2.

Cut M into two pieces, M_1 and M_2, separated by a hyperplane that is orthogonal to the cylindrical boundary of the prism. Rearrange the pieces M_i and re-paste them together along the original (and congruent) top and bottom faces. We are then left with a right cylinder C whose surrounding boundary is orthogonal to the new top and bottom faces. Since μ remains constant under this operation, it follows that

$$\mu(M) = \mu(M_1) + \mu(M_2) = \mu(C) = 0.$$

(Actually, such a cutting and rearrangement is possible only if the diameter of the top/bottom of M is sufficiently small compared with the height and angle of the cylindrical boundary; i.e. provided that M is not too 'fat'. If the base of M is too large, however, we can subdivide M into 'skinny' prisms by subdividing the top/bottom of M into convex bodies of sufficiently small diameter and considering separately each prism formed by taking the convex hull of the (disjoint) union of a piece of the bottom of M with its corresponding congruent piece of the top of M.)

Now let P be a convex polytope having facets $P_1, \ldots, P_m$, and corresponding outward unit normal vectors $u_1, \ldots, u_m$. Let $v \in \mathbf{R}^n$, and let $\bar{v}$ denote the straight line segment connecting the point v to the origin o. Without loss of generality, let us assume that $P_1, \ldots, P_j$ are exactly those facets of P such that $u_i \cdot v > 0$, for each $1 \le i \le j$. In this case, the Minkowski sum $P + \bar{v}$ can be expressed in the form

$$P + \bar{v} = P \cup \left(\bigcup_{i=1}^{j} (P_i + \bar{v}) \right),$$

where each term of the above union is either disjoint from the others, or intersects another in a convex body of dimension at most $n - 1$. It follows that

$$\mu(P + \bar{v}) = \mu(P) + \left(\sum_{i=1}^{j} \mu(P_i + \bar{v}) \right).$$

Notice, however, that each term of the form $P_i + \bar{v}$ is a prism, so that $\mu(P_i + \bar{v}) = 0$. Hence,

$$\mu(P + \bar{v}) = \mu(P), \tag{8.2}$$

for all convex polytopes P and all line segments $\bar{v}$.

By induction over finite Minkowski sums of line segments, it immediately follows from (8.2) that, for all convex polytopes P and all zonotopes Z,

$$\mu(Z) = 0, \quad \text{and} \quad \mu(P + Z) = \mu(P).$$

The continuity of μ then implies that

$$\mu(Y) = 0, \quad \text{and} \quad \mu(K + Y) = \mu(K), \tag{8.3}$$

for all $K \in \mathcal{K}^n$ and all zonoids Y.

Next, suppose that $K \in \mathcal{K}_c^n$ has a C^∞ support function h_K. By Proposition 8.1.2, there exist zonoids Y_1 and Y_2 such that $K + Y_2 = Y_1$. In this case, (8.3) implies that

$$\mu(K) = \mu(K + Y_2) = \mu(Y_1) = 0.$$

Since any centered convex body K can be approximated by a sequence K_i of C^∞ centered convex bodies, it follows (by continuity) that μ is zero on all of $\mathcal{K}_c^n$.

Now let Δ be an n-dimensional simplex, with one vertex at the origin. Let $u_1, \ldots, u_n$ denote the other vertices of Δ, and let P be the parallelotope spanned by the vectors $u_1, \ldots, u_n$. Let $v = u_1 + \cdots + u_n$. Let ξ_1 be the hyperplane passing through the points $u_1, \ldots, u_n$, and let ξ_2 be the hyperplane passing through the points $v - u_1, \ldots, v - u_n$. Finally, denote by P_* the set of all points of P lying between the hyperplanes ξ_1 and ξ_2. We can now write

$$P = \Delta \cup P_* \cup (-\Delta + v),$$

where each term of the union intersects another in dimension at most $n - 1$. Since P and P_* are centered, we have

$$0 = \mu(P) = \mu(\Delta) + \mu(P_*) + \mu(-\Delta + v) = \mu(\Delta) + \mu(-\Delta).$$

In other words, $\mu(\Delta) = -\mu(-\Delta)$. Meanwhile, we are given that $\mu(\Delta) = \mu(-\Delta)$. Therefore $\mu(\Delta) = 0$, for any simplex Δ.

Let P be a convex polytope in $\mathbf{R}^n$. The polytope P can be expressed as a finite union of simplices

$$P = \Delta_1 \cup \cdots \cup \Delta_m,$$

such that the intersection $\Delta_i \cap \Delta_j$ has dimension less than n, for all $i \neq j$. It follows that

$$\mu(P) = \sum_{i=1}^{m} \mu(\Delta_i) = 0.$$

Since the set of all convex polytopes is dense in $\mathcal{K}^n$, the continuity of μ then implies that $\mu(K) = 0$ for all $K \in \mathcal{K}^n$. $\qquad\square$

Theorem 8.1.4 is equivalent to the following theorem.

Theorem 8.1.5 (The volume characterization theorem) *Suppose that μ is a continuous translation invariant simple valuation on $\mathcal{K}^n$. Then there exists $c \in \mathbf{R}$ such that $\mu(K) + \mu(-K) = c\mu_n(K)$, for all $K \in \mathcal{K}^n$.*

Note that Theorem 8.1.5 implies that $\mu(K) = (c/2)\mu_n(K)$ for all centered convex bodies $K \in \mathcal{K}_c^n$.
Proof of equivalence Suppose that μ is a continuous translation invariant simple valuation on $\mathcal{K}^n$. For $K \in \mathcal{K}^n$, define

$$\nu(K) = \mu(K) + \mu(-K) - 2\mu([0,1]^n)\mu_n(K).$$

Then ν satisfies the hypotheses of Theorem 8.1.4, so that $\nu(K) = 0$ for all $K \in \mathcal{K}^n$. Therefore,

$$\mu(K) + \mu(-K) = c\mu_n(K),$$

where $c = 2\mu([0,1]^n)$. Hence, Theorem 8.1.4 implies Theorem 8.1.5. The reverse implication is obvious. $\qquad\square$

8.2 Even and odd valuations

A valuation μ on $\mathcal{K}^n$ is said to be *even* if

$$\mu(-K) = \mu(K)$$

for all $K \in \mathcal{K}^n$. If

$$\mu(-K) = -\mu(K)$$

for all $K \in \mathcal{K}^n$ then μ is said to be *odd*. Every valuation μ on $\mathcal{K}^n$ has a decomposition

$$\mu = \mu_{\text{even}} + \mu_{\text{odd}},$$

where μ_{even} is the even valuation defined by

$$\mu_{\text{even}}(K) = \tfrac{1}{2}(\mu(K) + \mu(-K))$$

and μ_{odd} is the odd valuation defined by

$$\mu_{\text{odd}}(K) = \tfrac{1}{2}(\mu(K) - \mu(-K)).$$

Theorem 8.1.5 implies that the volume μ_n is the only continuous, translation invariant, *even*, simple valuation on $\mathcal{K}^n$, up to a constant factor. More generally, for any continuous, translation invariant, simple valuation μ, there exists $c \in \mathbf{R}$ such that

$$\mu(K) = c\mu_n(K) + \mu_{\text{odd}}(K) \tag{8.4}$$

for all K.

A natural question at this point is that of whether the property of evenness is necessary to characterize volume. Do there exist any nontrivial continuous, translation invariant, *odd*, simple valuations? The answer turns out to be yes, even in the case of dimension 2. For example, let Δ denote the equilateral triangle in $\mathbf{R}^2$ of unit side length, centered at the origin and with a side parallel to the x-axis. Define a valuation η on $\mathcal{K}^2$ by the equation

$$\eta(K) = \mu_2(K + \Delta) - \mu_2(K + (-\Delta)). \tag{8.5}$$

It is clear that η is a translation invariant continuous function of K. To see that η is a valuation, we require the following proposition.

Proposition 8.2.1 *Suppose* $K, L, M \in \mathcal{K}^n$ *such that* $K \cup L$ *is convex. Then*

$$(K \cup L) + M = (K + M) \cup (L + M), \tag{8.6}$$

and

$$(K \cap L) + M = (K + M) \cap (L + M). \tag{8.7}$$

Proof The equation (8.6) is obvious, even when $K \cup L$ is not convex. To prove (8.7), suppose that $x \in (K \cap L) + M$. Then $x = y + m$, where $y \in K \cap L$ and $m \in M$. Evidently $x \in K + M$ and $x \in L + M$, so that

$$(K \cap L) + M \subseteq (K + M) \cap (L + M).$$

Next, suppose that $x \in (K + M) \cap (L + M)$. Then $x = a + m_1 = b + m_2$, where $a \in K$, $b \in L$, and $m_1, m_2 \in M$. Since $K \cup L$ is convex, with K and L each compact, the line segment with endpoints at a and b must contain a point $y \in K \cap L$, which we can express in the form $y = (1 - t)a + tb$ for some $0 \le t \le 1$. Let $m = (1 - t)m_1 + tm_2$. Since M is convex, $m \in M$, and $y + m \in (K \cap L) + M$. Meanwhile,

$$y + m = (1 - t)a + tb + (1 - t)m_1 + tm_2 = (1 - t)x + tx = x,$$

so that $x \in (K \cap L) + M$, and

$$(K \cap L) + M \supseteq (K + M) \cap (L + M).$$

This completes the proof. $\qquad\qquad\qquad\qquad\qquad\qquad\qquad\qquad\square$

Now suppose that $K, L \in \mathcal{K}^2$ such that $K \cup L$ is convex. If we set $M = \Delta$ then Proposition 8.2.1 implies that

$$
\begin{aligned}
\mu_2((K \cup L) + \Delta) &+ \mu_2((K \cap L) + \Delta) \\
&= \mu_2((K + \Delta) \cup (L + \Delta)) + \mu_2((K + \Delta) \cap (L + \Delta)) \\
&= \mu_2(K + \Delta) + \mu_2(L + \Delta),
\end{aligned}
$$

and similarly if Δ is replaced with $-\Delta$. Consequently, we have

$$\eta(K \cup L) + \eta(K \cap L) = \eta(K) + \eta(L).$$

Moreover, since μ_2 is even,

$$
\begin{aligned}
\eta(-K) &= \mu_2(-K + \Delta) - \mu_2(-K + (-\Delta)) = \mu_2(K + (-\Delta)) \\
&\quad\ -\mu_2(K + \Delta) \\
&= -\eta(K),
\end{aligned}
$$

so that η is an odd valuation on $\mathcal{K}^2$.

Fig. 8.3. The Minkowski sum $\Delta + (-\Delta)$.

It immediately follows that η is simple! To see this, suppose that K is a *symmetric* convex body; i.e., $K = -K$. Then we have

$$\eta(K) = \eta(-K) = -\eta(K),$$

so that $\eta(K) = 0$. However, all points and line segments are symmetric convex bodies, so that η must vanish in dimensions 0 and 1.

Finally, we check that $\eta \neq 0$. The fact that $\eta(\Delta) \neq 0$ follows easily from the *Brunn–Minkowski inequality* in the plane [85, p. 309]. However, this fact can also be seen by a simple and direct computation. It can be seen in Figure 8.3 that the Minkowski sum $\Delta + (-\Delta)$ is a regular hexagon of unit edge length, having area

$$\mu_2(\Delta + (-\Delta)) = 6\mu_2(\Delta),$$

whereas

$$\mu_2(\Delta + \Delta) = \mu_2(2\Delta) = 4\mu_2(\Delta).$$

Hence,

$$\eta(\Delta) = \mu_2(\Delta + \Delta) - \mu_2(\Delta + (-\Delta)) = -2\mu_2(\Delta) = -\frac{\sqrt{3}}{2} \neq 0.$$

Shortly after the discovery of Theorem 8.1.5, which characterizes the continuous translation invariant *even* simple valuations, Schneider used an analogous approach to characterize the continuous translation invariant *odd* simple valuations. This led to the following theorem.

Theorem 8.2.2 (Schneider's characterization theorem) *Suppose that μ is a continuous translation invariant odd simple valuation on $\mathcal{K}^n$. Then there exists a continuous odd function $g : \mathbf{S}^{n-1} \longrightarrow \mathbf{R}$ and a measure S_K on $\mathbf{S}^{n-1}$ such that*

$$\mu(K) = \int_{\mathbf{S}^{n-1}} g(u) \, \mathrm{d}S_K$$

for all $K \in \mathcal{K}^n$. □

Here $\mathbf{S}^{n-1}$ denotes the unit sphere in $\mathbf{R}^n$. The measure S_K is sometimes called the Aleksandrov–Fenchel–Jessen measure associated with K.

Since any continuous translation invariant simple valuation μ can be expressed as a sum $\mu = \mu_{\text{even}} + \mu_{\text{odd}}$ of even and odd valuations, Theorems 8.1.5 and 8.2.2 combine to improve the characterization (8.4). In other words, a continuous translation invariant simple valuation μ must have the form

$$\mu(K) = c\mu_n(K) + \int_{\mathbf{S}^{n-1}} g(u)\, \mathrm{d}S_K,$$

where $c \in \mathbf{R}$ and $g : \mathbf{S}^{n-1} \longrightarrow \mathbf{R}$ is a continuous odd function.

8.3 The volume theorem

In order to generalize Theorem 4.2.4 to the lattice Polycon(n), we require the following proposition relating rotation invariance to invariance under reflections.

Proposition 8.3.1 (Sah) *Let Δ be an n-dimensional simplex. There exist polytopes $P_1, \ldots, P_m$ such that*

$$\Delta = P_1 \cup \cdots \cup P_m,$$

where each term of this union intersects another in dimension at most $n-1$, and where each of the polytopes P_i is symmetric under a reflection across a hyperplane.

Proof Let $x_0, \ldots, x_n$ be the vertices of Δ, and let Δ_i be the facet of Δ opposite to x_i. Let z be the center of the inscribed sphere of Δ, and let z_i be the foot of the perpendicular from z to the facet Δ_i. For all $i < j$, let $A_{i,j}$ denote the convex hull of z, z_i, z_j, and the face $\Delta_i \cap \Delta_j$. Then

$$\Delta = \bigcup_{0 \le i < j \le n} A_{i,j},$$

where the distinct terms $A_{i,j}$ of this union intersect in at most dimension $n-1$. It is also evident that each $A_{i,j}$ is symmetric under reflection across the $n-1$ hyperplane determined by the point z and the face $\Delta_i \cap \Delta_j$. Now relabel the polytopes $A_{i,j}$ by a linear ordering $P_1, \ldots, P_m$, where $m = \frac{1}{2}n(n+1)$. This gives

$$\Delta = P_1 \cup \cdots \cup P_m,$$

where the polytopes P_i satisfy the desired conditions. $\qquad\square$

We now generalize Theorem 4.2.4 to the lattice Polycon(n).

Theorem 8.3.2 (The volume theorem for Polycon(n)**)** *Suppose that μ is a continuous rigid motion invariant simple valuation on $\mathcal{K}^n$ (or Polycon(n)). Then there exists $c \in \mathbf{R}$ such that $\mu(K) = c\mu_n(K)$, for all $K \in \mathcal{K}^n$ (or Polycon(n)).*

Recall from Groemer's extension Theorem 5.1.1 that a convex-continuous valuation is well defined on Polycon(n) if and only if it is well defined on the generating set $\mathcal{K}^n$.

Proof Since μ is translation invariant (as well as rotation invariant) and simple, Theorem 8.1.5 implies the existence of $a \in \mathbf{R}$ such that $\mu(K) + \mu(-K) = a\mu_n(K)$, for all $K \in \mathcal{K}^n$.

Let Δ be a simplex in $\mathbf{R}^n$. Then we have

$$\mu(\Delta) + \mu(-\Delta) = a\mu_n(\Delta). \tag{8.8}$$

If the dimension n of the ambient Euclidean space is even, then Δ differs from $-\Delta$ by a rotation, so that

$$\mu(\Delta) = \mu(-\Delta) = (a/2)\mu_n(\Delta).$$

Meanwhile, if n is odd, from Proposition 8.3.1 there exist polytopes $P_1, \ldots, P_m$ such that

$$\Delta = P_1 \cup \cdots \cup P_m,$$

where each term of this union intersects another in dimension at most $n-1$, and where each of the polytopes P_i is symmetric under a reflection across a hyperplane.

It follows that each P_i differs from $-P_i$ by a proper rigid motion (i.e. by a rotation followed by a translation), so that $\mu(-P_i) = \mu(P_i)$. Therefore,

$$\mu(-\Delta) = \sum_{i=1}^{m} \mu(-P_i) = \sum_{i=1}^{m} \mu(P_i) = \mu(\Delta). \tag{8.9}$$

Taken together, (8.8) and (8.9) imply that $\mu(\Delta) = (a/2)\mu_n(\Delta)$ for any simplex Δ.

Let $c = a/2$, and suppose that P is a convex polytope in $\mathbf{R}^n$. The polytope P can be expressed as a finite union of simplices

$$P = \Delta_1 \cup \cdots \cup \Delta_m,$$

such that the intersection $\Delta_i \cap \Delta_j$ has dimension less than n, for all $i \neq j$. It follows that

$$
\begin{aligned}
\mu(P) &= \mu(\Delta_1) + \cdots + \mu(\Delta_m) \\
&= c\mu_n(\Delta_1) + \cdots + c\mu_n(\Delta_m) \\
&= c\mu_n(P).
\end{aligned}
$$

Since the set of all convex polytopes is dense in $\mathcal{K}^n$, the continuity of μ then implies that $\mu(K) = c\mu_n(K)$ for all $K \in \mathcal{K}^n$. This concludes the proof of Theorem 8.3.2. □

8.4 The normalization of the intrinsic volumes

Recall from Theorem 4.2.2 that if P is a parallelotope in $\mathbf{R}^l$, and if we consider $\mathbf{R}^l \subseteq \mathbf{R}^n$ for some $n > l$, then $\mu_k^l(P) = \mu_k^n(P)$, for all $k \geq 0$. We will now show that this universal normalization holds when P is replaced by any polyconvex set.

Theorem 8.4.1 (The universal normalization theorem)

The valuations μ_i on $\mathrm{Polycon}(n)$ are normalized independently of the dimension n.

Proof Theorem 8.4.1 clearly holds for $i = 0$, since $\mu_0^n(K) = \mu_0(K) = 1$ for all non-empty compact convex subsets of $\mathbf{R}^n$ for all dimensions n. Moreover, it is clear that μ_k^k restricts to μ_k^l for all $l < k$, since both valuations are identically zero on polyconvex sets of dimension $l < k$.

Let $n > k$, and suppose that

$$
\mu_k^{n-1} \text{ restricts to } \mu_k^l \text{ for all } k \leq l \leq n-1. \tag{8.10}
$$

We need to show that the condition (8.10) holds for μ_k^n as well.

Since μ_k^n vanishes in dimension less than k, the restriction of μ_k^n to a k-dimensional plane in $\mathbf{R}^n$ is a continuous, invariant, simple valuation on $\mathrm{Polycon}(k)$. By the Volume Theorem 8.3.2 there exists $c \in \mathbf{R}$ such that $\mu_k^n(K) = c\mu_k^k(K)$ for all $K \in \mathrm{Polycon}(k)$. Since $\mu_k^n(P) = \mu_k^k(P)$ for all parallelotopes $P \in \mathrm{Par}(k)$ (by Theorem 4.2.2), it follows that $c = 1$ and $\mu_k^n = \mu_k^k$ on $\mathrm{Polycon}(k)$.

If $k = n - 1$ then we have achieved our goal. If $k < n - 1$, suppose that

$$
\mu_k^n \text{ restricts to } \mu_k^l \text{ for some } k \leq l < n-1. \tag{8.11}
$$

To complete the induction step and prove the theorem, we need to show that the condition (8.11) holds for μ_k^n and μ_k^{l+1} as well.

Denote by ν the restriction of μ_k^n to Polycon($l+1$). Then ν restricts to μ_k^l on Polycon(l) by the condition (8.11), whereas μ_k^{l+1} restricts to μ_k^l on Polycon(l) by the condition (8.10). It follows that $\nu - \mu_k^{l+1}$ vanishes on Polycon(l), so that $\nu - \mu_k^{l+1}$ is a continuous invariant simple valuation on Polycon($l+1$). By Theorem 8.3.2 there exists $c \in \mathbf{R}$ such that $\nu - \mu_k^{l+1} = c\mu_{l+1}^{l+1}$ on Polycon($l+1$). Since $\nu - \mu_k^{l+1}$ vanishes on all of Par($l+1$) by Theorem 4.2.2, we have $c = 0$ and $\nu = \mu_k^{l+1}$. The theorem now follows by double induction, first on l and then on n.   □

Henceforth we are justified in omitting reference to ambient spaces when discussing the intrinsic volumes, and so we shall simplify the notation μ_k^n to μ_k. The induction technique used in the proof of Theorem 8.4.1 actually leads to a much more powerful result, with which we shall begin Chapter 9.

8.5 Lattice points and volume

Let $\mathcal{B} = \{v_1, \ldots, v_n\}$ be a basis for $\mathbf{R}^n$, and let $\mathcal{L}$ denote the collection of points in $\mathbf{R}^n$ consisting of all vectors having integer coordinates with respect to $\mathcal{B}$; that is,

$$\mathcal{L} = \{a_1 v_1 + \cdots + a_n v_n \ : \ a_1, \ldots, a_n \in \mathbf{Z}\}.$$

Here $\mathbf{Z}$ denotes the set of integers.

The set $\mathcal{L}$ forms a discrete subgroup of $\mathbf{R}^n$ with respect to vector addition. Discrete subgroups of $\mathbf{R}^n$ are traditionally called *lattices*, and the elements of a lattice $\mathcal{L}$ are called *lattice points*. Lattices in $\mathbf{R}^n$ are not to be confused with the distributive lattices of sets treated in the preceding chapters – although the two notions are related, this relation will not be relevant in the present context. For the remainder of Chapter 8, the term *lattice* will refer to discrete additive subgroups of $\mathbf{R}^n$.

In this section we ask the following question. If a convex body K is moved in $\mathbf{R}^n$ by a random Euclidean motion, what is the expected number of lattice points to be found in K? In other words, what is the expectation of the random variable $|gK \cap \mathcal{L}|$, where g is a random Euclidean motion? Using Theorem 8.1.1 we will show that this expected number is proportional to the volume of K.

Let C denote the parallelotope

$$C = \{a_1 v_1 + \cdots + a_n v_n \ : \ 0 \le a_1, \ldots, a_n \le 1\}.$$

The parallelotope C is called a *fundamental domain* for the lattice $\mathcal{L}$. If A is a matrix whose ith column is given by the coordinates of the vector v_i (with respect to the standard orthonormal basis for $\mathbf{R}^n$), then the volume of C is given by $\mu_n(C) = |\det A|$.

Proposition 8.5.1 *For any $x \in \mathbf{R}^n$ and any positive integer k, the translate $kC + x$ of kC contains at least k^n points of $\mathcal{L}$ and at most $(k+1)^n$ points of $\mathcal{L}$.*

Proof For $x = x_1 v_1 + \cdots + x_n v_n$ the set $(kC + x) \cap \mathcal{L}$ consists of all vectors $v = y_1 v_1 + \cdots + y_n v_n$ such that $x_i \le y_i \le k + x_i$. There are either k or $k + 1$ possible integer values for y_i within this interval, for each $i \in \{1, \ldots, n\}$. Therefore $kC + x$ contains between k^n and $(k+1)^n$ points of $\mathcal{L}$. $\qquad\square$

We consider next a simplified version of our original question. Suppose that a polyconvex set K is moved in $\mathbf{R}^n$ by a random translation x, resulting in a new body $K + x$. What is the expected number of lattice points in $K + x$?

Theorem 8.5.2 *Let $K \in \mathrm{Polycon}(n)$, and let X_K denote the number of points in the set $(K + x) \cap \mathcal{L}$, for a random translation x. Then the expectation $E(X_K)$ is given by*

$$E(X_K) = \frac{\mu_n(K)}{\mu_n(C)}.$$

Proof To begin, note that for $x \in \mathbf{R}^n$ the number of lattice points in $K + x$ is given by $\mu_0((K + x) \cap \mathcal{L})$, where μ_0 is the Euler characteristic. In order to compute the expectation $E(X_K)$ we must average over all $x \in \mathbf{R}^n$ the function $\mu_0(gK \cap \mathcal{L})$. This makes no sense, however, since such an integral would diverge. Nonetheless, because $\mathcal{L}$ is symmetric under the set of translations by points of $\mathcal{L}$, it is sufficient to average over the set of translations by vectors $x \in C$. Thus we have

$$E(X_K) = \int_C \mu_0((K + x) \cap \mathcal{L})\, dx,$$

so that $E(X_K)$ is a translation invariant monotonically increasing valuation in the parameter K. Evidently $E(X_K) = 0$ if K has dimension less than n, so the valuation $E(X_K)$ is also simple. It then follows from

Theorem 8.1.1 that there exists $\alpha \in \mathbf{R}$ such that

$$E(X_K) = \alpha \mu_n(K), \tag{8.12}$$

for all $K \in \text{Polycon}(n)$.

To compute the constant α, consider $K = C$. For all positive integers k, Proposition 8.5.1 asserts that

$$k^n \le \mu_0((kC + x) \cap \mathcal{L}) \le (k+1)^n.$$

It then follows that

$$k^n \le E(X_{kC}) \le (k+1)^n.$$

From (8.12) we then obtain

$$k^n \le \alpha k^n \mu_n(C) \le (k+1)^n,$$

for all positive integers k. It follows that $\alpha = 1/\mu_n(C)$. □

We are now able to answer the question of how many lattice points are expected inside gK, for a random Euclidean motion g.

Theorem 8.5.3 *Let $K \in \text{Polycon}(n)$, and let X_K denote the number of points in the set $gK \cap \mathcal{L}$, for a random Euclidean motion g. Then the expectation $E(X_K)$ is given by*

$$E(X_K) = \frac{\mu_n(K)}{\mu_n(C)}.$$

Proof To begin, note that, for $g \in E_n$, the number of lattice points in gK is given by $\mu_0(gK \cap \mathcal{L})$, where μ_0 is the Euler characteristic. In order to compute the expectation $E(X_K)$ we must average the function $\mu_0(gK \cap \mathcal{L})$ over all motions g. Again this appears to make no sense, since the required integral would diverge. However, because $\mathcal{L}$ is symmetric under the set of translations by points of $\mathcal{L}$, it is sufficient to average over the set of motions $\widetilde{E}_n$ consisting of any rotation or reflection $\phi \in O(n)$ followed by a translation by a vector $x \in C$. Hence, we have

$$\begin{aligned} E(X_K) &= \int_{\widetilde{E}_n} \mu_0(gK \cap \mathcal{L}) \, dg \\ &= \int_{O(n)} \int_C \mu_0((\phi K + x) \cap \mathcal{L}) \, dx \, d\phi, \end{aligned}$$

where the integrals are taken with respect to the unique invariant probability measures on C and $O(n)$. Since

$$\int_C \mu_0((\phi K + x) \cap \mathcal{L})\, \mathrm{d}x = \frac{\mu_n(\phi K)}{\mu_n(C)} = \frac{\mu_n(K)}{\mu_n(C)}$$

by Theorem 8.5.2, it follows that

$$E(X_K) = \int_{O(n)} \frac{\mu_n(K)}{\mu_n(C)}\, \mathrm{d}\phi = \frac{\mu_n(K)}{\mu_n(C)}$$

as well. □

8.6 Remarks on Hilbert's third problem

Let P be a polytope in $\mathbf{R}^n$. A *dissection* of P is an expression of P as a union

$$P = P_1 \cup P_2 \cup \cdots \cup P_m$$

of polytopes $P_1, \ldots, P_m$ such that each intersection $P_i \cap P_j$ has dimension less than n for all $i \neq j$.

Two polytopes P and Q in $\mathbf{R}^n$ are said to be *scissors congruent* if there exist dissections of P and Q into a finite set of polytopes $P_1, P_2, \ldots, P_m$ and $Q_1, Q_2, \ldots, Q_m$ respectively, such that each P_i is congruent to Q_i by some rigid motion of $\mathbf{R}^n$; that is, such that for each i we have $P_i = g_i Q_i$ for some rigid motion g_i of $\mathbf{R}^n$. It is clear that, if P and Q are scissors congruent, then they have the same volume: $\mu_n(P) = \mu_n(Q)$.

At the Paris International Congress of Mathematicians in 1900, David Hilbert posed the converse question (see [44]): *if two polytopes P and Q in $\mathbf{R}^n$ have the same volume, are they necessarily scissors congruent?* This question, along with subsequent variations, has come to be known as *Hilbert's third problem*.

It can easily be shown that the answer is 'yes' for polytopes in the plane; that is, if $n = 2$ (see [81, p. 5]). For dimensions $n \geq 3$, however, a *negative* answer was given by Dehn in [19] only a year after Hilbert had posed the problem (see also [8, 81]).

Dehn discovered a functional ψ on the set of polytopes in $\mathbf{R}^n$ (for $n \geq 3$), known as the *Dehn invariant*, such that $\psi(P) = \psi(Q)$ whenever P and Q are scissors congruent. He then exhibited two polytopes P and Q having the same volume such that $\psi(P) \neq \psi(Q)$.

In modern parlance the term 'Dehn invariant' actually refers to a family of scissors congruence invariant functionals that distinguish between some polytopes of the same volume. An example in $\mathbf{R}^3$ is constructed as follows. Let $\alpha = (\arccos(1/3))/\pi$, and note that α is irrational (see, for example, [8, p. 102]). Recall again that $\mathbf{R}$ is a vector space of infinite dimension over the field $\mathbf{Q}$ of rational numbers. Let $f : \mathbf{R} \longrightarrow \mathbf{R}$ be a $\mathbf{Q}$-linear functional on $\mathbf{R}$ such that $f(1) = 0$ and $f(\alpha) = 1$. (Such a function is possible because α is irrational.) For a compact convex polyhedron P in $\mathbf{R}^3$, let $M_1, \ldots, M_m$ denote the edges of P, and let $\theta_1, \ldots, \theta_m$ denote the outer dihedral angles between the facets adjacent to each corresponding edge. Define

$$\psi(P) = \sum_{i=1}^{m} \mu_1(M_i) f \left(\frac{\theta_i}{\pi} \right).$$

With some effort one verifies that ψ is invariant under scissors congruence. Let Δ denote the regular tetrahedron in $\mathbf{R}^3$ having unit volume. Since the dihedral angle between any two adjacent facets of Δ is given by $\arccos(1/3)$, we have $\psi(\Delta) > 0$. Meanwhile, the unit cube C in $\mathbf{R}^3$ has outer dihedral angles $3\pi/2$, so that $\psi(C) = 0$. It follows that C and Δ are not scissors congruent, in spite of having the same volume.

A more general form of Dehn invariant is given in terms of tensor products (see also [81, p. 2]). Suppose $n \geq 3$, and let P be a compact convex polytope in $\mathbf{R}^n$ having $(n-2)$-dimensional faces $M_1, \ldots, M_m$. Let $\theta_1, \ldots, \theta_m$ denote the angles between the facets adjacent to each corresponding $(n-2)$-face. Denote by $\mathbf{R}/\mathbf{Z}$ the normalized circle group, and let $\mathbf{R} \otimes_{\mathbf{Z}} \mathbf{R}/\mathbf{Z}$ denote the tensor product of the Abelian groups $\mathbf{R}$ and $\mathbf{R}/\mathbf{Z}$. Define a functional $\Psi : \mathcal{P}^n \longrightarrow \mathbf{R} \otimes_{\mathbf{Z}} \mathbf{R}/\mathbf{Z}$ by

$$\Psi(P) = \sum_{i=1}^{m} \mu_{n-2}(M_i) \otimes_{\mathbf{Z}} \left(\frac{\theta_i}{\pi} \right).$$

The functional Ψ is a Dehn invariant for all $n \geq 3$.

Hilbert's original question has many variations. Let G be any group of rigid motions of $\mathbf{R}^n$, for example, the group of translations, or some finite group of symmetries. We say that two polytopes P and Q are *scissors congruent with respect to G* if there exist dissections of P and Q into a finite set of polytopes $P_1, P_2, \ldots, P_m$ and $Q_1, Q_2, \ldots, Q_m$ respectively, so that each P_i is congruent to Q_i by some motion in G; that is, so that for each i we have $P_i = g_i Q_i$ for some $g_i \in G$. Once again it is clear that, if P and Q are scissors congruent with respect to G, then $\mu_n(P) = \mu_n(Q)$. Once again one may pose the converse question. The answer remains in the negative for $n \geq 3$, but, for the case $n = 2$ (the Euclidean plane), the answer is more interesting. In particular, suppose that $G = T_2$, the set of all translations of the plane. In this case Hilbert's question has once again a negative answer, even in the plane! This follows from our construction (8.5) of the valuation η in Section 8.2. Recall that η was defined by

$$\eta(K) = \mu_2(K + \Delta) - \mu_2(K + (-\Delta)),$$

where $K \in \mathcal{K}^2$ and Δ is the equilateral triangle of unit side length, centered at the origin. The valuation η is simple and translation invariant. It follows that, if P and Q are polygons in $\mathbf{R}^2$ that are scissors congruent with respect to T_2, then $\eta(P) = \eta(Q)$. However, recall also that $\eta(K) = 0$ whenever K is a centered convex body (such as a rectangle), while $\eta(\Delta) \neq 0$. Let $P = \Delta$, and let Q be a square with the same area as P. Since $\eta(P) \neq \eta(Q)$, the polygons P and Q cannot be scissors congruent with respect to the translation group T_2.

In view of Theorem 8.1.5, one might suspect that the answer to Hilbert's third problem is once again 'Yes' in dimension 2, provided that we allow the group G to contain both the translations of the plane and the reflection through the origin (corresponding to scalar multiplication by -1). Indeed, this was shown to be true by Hadwiger and Glur in [40]. A detailed treatment of the results of Hadwiger and Glur may also be found in [8, pp. 69–92] and [81].

The remarks in this section give only a hint of the beauty and complexity to be found in the study of polytopes, dissections, and congruences. For a more thorough treatment of this theory, see [8, 71, 81].

8.7 Notes

Theorem 8.3.2 is due to Hadwiger, who gave a long and difficult proof in [39]. The more general Theorem 8.1.5 is due to Klain [50]. Schneider's characterization theorem appeared in [86]. Proposition 8.3.1 is due to Sah [81, pp. 16–17]. For a discussion of spherical harmonics and other background for Proposition 8.1.2, see [85, p. 184] and [33, 34].

Theorem 8.1.5 leads in turn to a connection between continuous even valuations on compact convex sets and continuous functions on Grassmannians. This connection can be described in part by using generating distributions for symmetric compact convex sets. For a detailed discussion, see [54]. The question of how to characterize volume as a valuation on defined only on polytopes remains open (see [72]).

9

Hadwiger's characterization theorem

In Section 9.1 we use the volume Theorem 8.3.2 to complete our characterization of invariant valuations on polyconvex sets with one of the most beautiful and important theorems in geometric probability, Hadwiger's characterization theorem. We then use Hadwiger's theorem to derive simple proofs of numerous results in integral geometry and geometric probability. In Sections 9.3 and 9.4 we give cleaner re-statements of the intrinsic volume formulas of Chapter 6. Sections 9.2, 9.7, and 9.8 deal with the computation of intrinsic volumes in special cases. In Section 9.6 the mean projection formula of Section 9.4 is combined with the results of Section 8.5 to yield a generalization of the Buffon needle problem to spaces and planes of arbitrary finite dimension.

9.1 A proof of Hadwiger's characterization theorem

The following result generalizes Theorem 4.2.5 to the lattice Polycon(n). As in the case for parallelotopes, this theorem is, in fact, equivalent to the associated volume Theorem 8.3.2.

Theorem 9.1.1 (Hadwiger's characterization theorem) *The valuations μ_0, μ_1, ..., μ_n form a basis for the vector space of all convex-continuous rigid motion invariant valuations defined on polyconvex sets in $\mathbf{R}^n$.*

Proof Let μ be a convex-continuous invariant valuation on Polycon(n). Let H be a hyperplane in $\mathbf{R}^n$; that is, a linear variety of dimension $n-1$. The restriction of μ to H is an invariant valuation on H. Proceeding by induction, we may assume that

$$\mu(A) = \sum_{i=0}^{n-1} c_i \mu_i(A),$$

for every polyconvex $A \subseteq H$. Thus, the valuation

$$\mu - \sum_{i=0}^{n-1} c_i \mu_i$$

vanishes on all lower dimensional polyconvex sets of $\mathbf{R}^n$. This follows from the invariance of the valuations μ_i and from the fact that any lower dimensional polyconvex set is contained in the image of some rigid motion of the hyperplane H. By the volume Theorem 8.3.2,

$$\mu - \sum_{i=0}^{n-1} c_i \mu_i = c_n \mu_n,$$

where μ_n is the volume on $\mathbf{R}^n$, and where c_n is a real constant. In other words,

$$\mu = \sum_{i=0}^{n} c_i \mu_i.$$

$\square$

A valuation μ on $\mathrm{Polycon}(n)$ is said to be *homogeneous* of degree $k > 0$ if

$$\mu(\alpha K) = \alpha^k \mu(K)$$

for all $K \in \mathrm{Polycon}(n)$ and all $\alpha \geq 0$.

Corollary 9.1.2 *Let μ be a convex-continuous rigid motion invariant valuation defined on* $\mathrm{Polycon}(n)$ *that is also homogeneous of degree k, for some $0 \leq k \leq n$. Then there exists $c \in \mathbf{R}$ such that $\mu(K) = c\mu_k(K)$ for all $K \in \mathrm{Polycon}(n)$.*

Proof By Theorem 9.1.1 there exist $c_1, \ldots, c_n \in \mathbf{R}$ such that

$$\mu = \sum_{i=0}^{n} c_i \mu_i.$$

If $P = [0,1]^n$ then, for $\alpha > 0$,

$$\mu(\alpha P) = \sum_{i=0}^{n} c_i \mu_i(\alpha P) = \sum_{i=0}^{n} c_i \mu_i(P) \alpha^i = \sum_{i=0}^{n} \binom{n}{i} c_i \alpha^i.$$

Meanwhile,

$$\mu(\alpha P) = \alpha^k \mu(P) = \sum_{i=0}^{n} c_i \mu_i(P) \alpha^k = \sum_{i=0}^{n} \binom{n}{i} c_i \alpha^k.$$

Therefore, $c_i = 0$ if $i \neq k$, and $\mu = c_k \mu_k$.   □

9.2 The intrinsic volumes of the unit ball

As an application of Theorem 9.1.1, we compute the intrinsic volumes $\mu_i(B_n)$ of the unit ball in $\mathbf{R}^n$. We require the following facts about Minkowski sums.

Let K, L be compact convex sets, and let $\alpha \geq 0$. Recall that the Minkowski sum $K + \alpha L$ is defined by

$$K + \alpha L = \{x + \alpha y : x \in K \text{ and } y \in L\}.$$

Proposition 9.2.1 *For $K \in \mathcal{K}^n$ and any unit vector $u \in \mathbf{R}^n$,*

$$\mu_n(K + \epsilon \bar{u}) = \mu_n(K) + \epsilon \mu_{n-1}(K|u^\perp).$$

Here $\bar{u}$ denotes the straight line segment connecting the point u to the origin o.

Proof Let $L = K + \epsilon \bar{u}$. The volume μ_n of L can be computed by integrating over the hyperplane $u^\perp$ the length of each linear slice of L; that is,

$$\mu_n(L) = \int_{u^\perp} \mu_1(L \cap \ell_x) \, dx,$$

where ℓ_x denotes the straight line parallel to u through the point $x \in u^\perp$. Since $\mu_1(L \cap \ell_x) = \mu_1(K \cap \ell_x) + \epsilon$ for all $x \in K|u^\perp$, while $L \cap \ell_x = \emptyset$ if $x \notin K|u^\perp$, we have

$$\begin{aligned}
\mu_n(L) &= \int_{u^\perp} \mu_1(L \cap \ell_x) \, dx \\
&= \int_{K|u^\perp} \mu_1(K \cap \ell_x) + \epsilon \, dx \\
&= \mu_n(K) + \epsilon \mu_{n-1}(K|u^\perp).
\end{aligned}$$

                □

Let C_n denote the n-dimensional unit cube. Recall from Theorem 4.2.1 that

$$\mu_i(C_n) = \binom{n}{i},$$

for $0 \leq i \leq n$.

Proposition 9.2.2 *For $\epsilon \geq 0$,*

$$\mu_n(C_n + \epsilon B_n) = \sum_{i=0}^{n} \mu_i(C_n)\omega_{n-i}\epsilon^{n-i} = \sum_{i=0}^{n} \binom{n}{i}\omega_{n-i}\epsilon^{n-i}. \qquad (9.1)$$

Proof Let $u_1, u_2, \ldots, u_n$ denote the standard orthonormal basis for $\mathbf{R}^n$. Let $\bar{u}_i$ denote the line segment with endpoints at the origin o and the point u_i.

From Proposition 9.2.1 we have

$$\mu_n(B_n + \epsilon\bar{u}_1) = \omega_n + \epsilon\omega_{n-1} = \sum_{i=0}^{1} \binom{1}{i}\omega_{n-i}\epsilon^i,$$

for all $n \geq 1$.

Suppose that the equation (9.1) holds for lower dimensions, and that

$$\mu_n(B_n + \epsilon\bar{u}_1 + \cdots + \epsilon\bar{u}_k) = \sum_{i=0}^{k} \binom{k}{i}\omega_{n-i}\epsilon^i,$$

for some $1 \leq k < n$. Then

$$\mu_n(B_n + \epsilon\bar{u}_1 + \cdots + \epsilon\bar{u}_{k+1})$$
$$= \mu_n(B_n + \epsilon\bar{u}_1 + \cdots + \epsilon\bar{u}_k) + \epsilon\mu_{n-1}(B_n + \epsilon\bar{u}_1 + \cdots + \epsilon\bar{u}_k | u_{k+1}^{\perp})$$
$$= \sum_{i=0}^{k} \binom{k}{i}\omega_{n-i}\epsilon^i + \epsilon\mu_{n-1}(B_{n-1} + \epsilon\bar{u}_1 + \cdots + \epsilon\bar{u}_k)$$
$$= \sum_{i=0}^{k} \binom{k}{i}\omega_{n-i}\epsilon^i + \sum_{i=0}^{k} \binom{k}{i}\omega_{n-1-i}\epsilon^{i+1}$$
$$= \sum_{i=0}^{k+1} \left(\binom{k}{i} + \binom{k}{i-1}\right)\omega_{n-i}\epsilon^i$$
$$= \sum_{i=0}^{k+1} \binom{k+1}{i}\omega_{n-i}\epsilon^i.$$

Since

$$C_n = \overline{u}_1 + \cdots + \overline{u}_n,$$

the equation (9.1) follows by induction to the step $k + 1 = n$. $\qquad \square$

Theorem 9.2.3 (Steiner's formula) *For $K \in \mathcal{K}^n$ and $\epsilon \geq 0$,*

$$\mu_n(K + \epsilon B_n) = \sum_{i=0}^{n} \mu_i(K)\omega_{n-i}\epsilon^{n-i}. \qquad (9.2)$$

Proof Let $\eta(K) = \mu_n(K + B_n)$, for $K \in \mathcal{K}^n$. It follows from Proposition 8.2.1 that η is a continuous invariant valuation. Theorem 9.1.1 then implies the existence of constants $c_0, \ldots, c_n \in \mathbf{R}$, such that

$$\eta(K) = \sum_{i=0}^{n} c_i \mu_i(K),$$

for all $K \in \mathcal{K}^n$. Therefore, for $\epsilon > 0$, we have

$$
\begin{aligned}
\mu_n(K + \epsilon B_n) &= \epsilon^n \mu_n\left(\frac{1}{\epsilon}K + B_n\right) = \epsilon^n \sum_{i=0}^{n} c_i \mu_i(K)\frac{1}{\epsilon^i} \\
&= \sum_{i=0}^{n} c_i \mu_i(K)\epsilon^{n-i}. \qquad (9.3)
\end{aligned}
$$

Setting $K = C_n$ and comparing equations (9.1) and (9.3) we find that $c_i = \omega_{n-i}$. $\qquad \square$

We are now able to compute the intrinsic volumes $\mu_i(B_n)$.

Theorem 9.2.4 (The intrinsic volumes of the unit ball) *For $0 \leq i \leq n$,*

$$\mu_i(B_n) = \binom{n}{i}\frac{\omega_n}{\omega_{n-i}} = \begin{bmatrix} n \\ i \end{bmatrix}\omega_i.$$

Proof Setting $K = B_n$ in (9.2), we obtain

$$
\begin{aligned}
\sum_{i=0}^{n} \mu_i(B_n)\omega_{n-i}\epsilon^{n-i} &= \mu_n(B_n + \epsilon B_n) = (1 + \epsilon)^n \mu_n(B_n) \\
&= \sum_{i=0}^{n} \binom{n}{i}\omega_n \epsilon^{n-i},
\end{aligned}
$$

for all $\epsilon > 0$. The proposition then follows from a comparison of the coefficients of each ϵ^{n-i}. $\qquad \square$

9.3 Crofton's formula

We are now ready to compute the constants C_k^n from Theorem 7.2.1. Specifically, we have the following remarkable result.

Theorem 9.3.1 *For* $0 \leq k \leq n$ *and* $K \in \mathcal{K}^n$,

$$\mu_{n-k}(K) = \lambda_k^n(\mathrm{Graff}(K; k)). \tag{9.4}$$

In other words, $C_k^n = 1$ for all $n, k \geq 0$. We shall see that our judicious choice of normalization (6.1) for $[n]$ leads in turn to this expression of the equation (9.4), free from normalizing constant factors. Once again, note that the equation (9.4) is asserted only for the case in which K is *convex* and *not* for arbitrary polyconvex sets.

Proof Recall from Section 7.2 that

$$C_k^n \lambda_k^n(\mathrm{Graff}(K; k)) = \mu_{n-k}(K)$$

for all compact convex sets K. For the case $K = B_n$, we have

$$C_k^n \lambda_k^n(\mathrm{Graff}(B_n; k)) = \mu_{n-k}(B_n) = \binom{n}{n-k}\frac{\omega_n}{\omega_k} = \omega_{n-k}\begin{bmatrix}n\\k\end{bmatrix}.$$

Meanwhile,

$$
\begin{aligned}
\lambda_k^n(\mathrm{Graff}(B_n; k)) &= \int_{\mathrm{Gr}(n,k)} \int_{V_0^\perp} \mu_0(B_n \cap (V_0 + p)) \, dp \, d\nu_k^n(V_0) \\
&= \int_{\mathrm{Gr}(n,k)} \int_{V_0^\perp} I_{B_{n-k}} \, dp \, d\nu_k^n(V_0) \\
&= \int_{\mathrm{Gr}(n,k)} \omega_{n-k} \, d\nu_k^n(V_0) \\
&= \omega_{n-k}\nu_k^n(\mathrm{Gr}(n,k)) \\
&= \omega_{n-k}\begin{bmatrix}n\\k\end{bmatrix}.
\end{aligned}
$$

Hence, we have

$$C_k^n \omega_{n-k}\begin{bmatrix}n\\k\end{bmatrix} = \omega_{n-k}\begin{bmatrix}n\\k\end{bmatrix},$$

so that $C_k^n = 1$. $\qquad\square$

Theorem 9.3.1 enables us in turn to rewrite Hadwiger's formula (7.8):

$$\mu_{n-k}(K) = \int_{\mathrm{Graff}(n,k)} \mu_0(K \cap V) \, d\lambda_k^n(V). \tag{9.5}$$

Unlike (9.4), the formula (9.5) is valid for all polyconvex sets K, although it reduces to (9.4) in the event that K is actually convex. In a similar vein, we obtain the following relation between intrinsic volumes of different degree, thereby generalizing Theorem 9.3.1.

Theorem 9.3.2 (Crofton's formula) *For $0 \le i, j \le n$ and $K \in$* Polycon(n),

$$\int\limits_{\text{Graff}(n,n-i)} \mu_j(K \cap V)\, d\lambda^n_{n-i}(V) = \begin{bmatrix} i+j \\ j \end{bmatrix} \mu_{i+j}(K). \qquad (9.6)$$

Note that, if $j = 0$ and $K \in \mathcal{K}^n$, then the identity (9.6) reduces to (9.4). *Proof* If $i + j > n$ then both sides of (9.6) are zero. Suppose that $i + j \le n$. For $K \in \mathcal{K}^n$, define

$$\eta(K) = \int\limits_{\text{Graff}(n,n-i)} \mu_j(K \cap V)\, d\lambda^n_{n-i}(V).$$

On applying (9.5) we then obtain

$$\eta(K) = \int\limits_{\text{Graff}(n,n-i)} \int\limits_{\text{Graff}(V,n-i-j)} \mu_0(K \cap V \cap W)\, d\lambda^{n-i}_{n-i-j}(W)\, d\lambda^n_{n-i}(V),$$

where Graff$(V, n - i - j)$ denotes the space of linear varieties $W \subseteq V$ having dimension $n - i - j$. Since each $W \subseteq V$, we can rewrite the previous integral as

$$\int_{\text{Graff}(n,n-i)} \int_{\text{Graff}(V,n-i-j)} \mu_0(K \cap W)\, d\lambda^{n-i}_{n-i-j}(W)\, d\lambda^n_{n-i}(V)$$

$$= \int_{V_0 \in \text{Gr}(n,n-i)} \int_{V_0^\perp} \int_{W_0 \in \text{Gr}(V_0,n-i-j)} \int_{W_0^\perp \cap V_0}$$
$$\mu_0(K \cap (W_0 + p + q))\, dq\, d\nu^{n-i}_{n-i-j}(W_0)\, dp\, d\nu^n_{n-i}(V_0)$$

$$= \int_{\text{Gr}(n,n-i)} \int_{\text{Gr}(V_0,n-i-j)} \int_{V_0^\perp} \int_{W_0^\perp \cap V_0}$$
$$\mu_0(K \cap (W_0 + p + q))\, dq\, dp\, d\nu^{n-i}_{n-i-j}(W_0)\, d\nu^n_{n-i}(V_0)$$

$$= \int_{\text{Gr}(n,n-i)} \int_{\text{Gr}(V_0,n-i-j)} \int_{V_0^\perp \oplus (W_0^\perp \cap V_0)}$$
$$\mu_0(K \cap (W_0 + v))\, dv\, d\nu^{n-i}_{n-i-j}(W_0)\, d\nu^n_{n-i}(V_0)$$

$$= \int_{\mathrm{Gr}(n,n-i)} \int_{\mathrm{Gr}(V_0,n-i-j)} \int_{W_0^\perp}$$
$$\mu_0(K \cap (W_0 + v)) \, dv \, d\nu_{n-i-j}^{n-i}(W_0) \, d\nu_{n-i}^n(V_0)$$

$$= \int_{\mathrm{Gr}(n,n-i)} \int_{\mathrm{Gr}(V_0,n-i-j)} \int_{W_0^\perp} I_{K|W_0^\perp} \, dv \, d\nu_{n-i-j}^{n-i}(W_0) \, d\nu_{n-i}^n(V_0)$$

$$= \int_{\mathrm{Gr}(n,n-i)} \int_{\mathrm{Gr}(V_0,n-i-j)} \mu_{i+j}(K|W_0^\perp) \, d\nu_{n-i-j}^{n-i}(W_0) \, d\nu_{n-i}^n(V_0),$$

since $\dim(W_0^\perp) = i + j$. Because μ_{i+j} is a continuous valuation, homogeneous of degree $i + j$, the linearity of the integrals above implies that η is a continuous valuation on $\mathcal{K}^n$, homogeneous of degree $i + j$. It then follows from the invariance of μ_{i+j} and of the measures ν_{n-i-j}^{n-i} and ν_{n-i}^n that η is invariant. By Corollary 9.1.2, there exists $c \in \mathbf{R}$ such that $\eta = c\mu_{i+j}$. To compute the constant c, set $K = B_n$, the unit ball in $\mathbf{R}^n$, and note that

$$\eta(B_n) \;=\; \int_{\mathrm{Gr}(n,n-i)} \int_{\mathrm{Gr}(V_0,n-i-j)} \mu_{i+j}(B_n|W_0^\perp) \, d\nu_{n-i-j}^{n-i}(W_0) \, d\nu_{n-i}^n(V_0)$$

$$=\; \begin{bmatrix} n \\ n-i \end{bmatrix} \begin{bmatrix} n-i \\ n-i-j \end{bmatrix} \omega_{i+j},$$

while

$$c\mu_{i+j}(B_n) = c \begin{bmatrix} n \\ i+j \end{bmatrix} \omega_{i+j},$$

by Theorem 9.2.4. Hence,

$$c = \begin{bmatrix} n \\ i+j \end{bmatrix}^{-1} \begin{bmatrix} n \\ n-i \end{bmatrix} \begin{bmatrix} n-i \\ n-i-j \end{bmatrix} = \begin{bmatrix} i+j \\ j \end{bmatrix}.$$

□

9.4 The mean projection formula revisited

The conclusion of Theorem 9.3.1 that $C_k^n = 1$ for all $0 \le k \le n$ also allows us to re-state Theorem 7.4.1, thereby generalizing Cauchy's formula 5.5.2.

Theorem 9.4.1 (The mean projection formula) *For* $0 \le k \le n$ *and* $K \in \mathcal{K}^n$,

$$\mu_k(K) = \int_{\mathrm{Gr}(n,k)} \mu_k(K|V_0) \, d\nu_k^n(V_0). \qquad (9.7)$$

□

In other words, the kth intrinsic volume $\mu_k(K)$ of a compact convex subset K of any dimension in $\mathbf{R}^n$ is *equal* to the integral of the k-volumes of the projections of K onto all k-dimensional subspaces of $\mathbf{R}^n$.

Recall that the intrinsic volumes μ_k are normalized; that is, $\mu_k(L)$ of an l-dimensional convex body L is the same regardless of the dimension $n \geq l$ of the ambient space $\mathbf{R}^n$. The absence of any additional normalizing factor in (9.7) demonstrates once again the importance of making the correct choice of normalization for the Grassmannian measures ν_k^n.

The mean projection formula can also be expressed in probabilistic terms, using random variables. For a compact convex set K in $\mathbf{R}^n$ let $X_k(K)$ denote the k-volume of a projection of K onto a *randomly chosen* k-dimensional subspace $V \in \mathrm{Gr}(n,k)$. The expectation $E(X_k(K))$ is computed by averaging over all projections; that is, by integrating over all subspaces with respect to the Haar *probability* measure on $\mathrm{Gr}(n,k)$, to wit,

$$E(X_k(K)) = \int_{\mathrm{Gr}(n,k)} \mu_k(K|V) \, \mathrm{d}V,$$

where

$$\int_{\mathrm{Gr}(n,k)} \mathrm{d}V = 1.$$

Integrating instead with respect to the measure ν_k^n, we have

$$E(X_k(K)) = \begin{bmatrix} n \\ k \end{bmatrix}^{-1} \int_{\mathrm{Gr}(n,k)} \mu_k(K|V) \, \mathrm{d}\nu_k^n.$$

Thus, we obtain the following version of Theorem 9.4.1.

Corollary 9.4.2 *For* $0 \leq k \leq n$ *and* $K \in \mathcal{K}^n$,

$$\mu_k(K) = \begin{bmatrix} n \\ k \end{bmatrix} E(X_k(K)).$$

$\square$

Hadwiger's Theorem 9.1.1 yields a simple proof of a yet more general form of Theorem 9.4.1.

Theorem 9.4.3 (Kubota's theorem) *For* $0 \leq k \leq l \leq n$ *and* $K \in \mathcal{K}^n$,

$$\int_{\mathrm{Gr}(n,l)} \mu_k(K|V) \, \mathrm{d}\nu_l^n(V) = \begin{bmatrix} n-k \\ l-k \end{bmatrix} \mu_k(K).$$

Proof Define a valuation η on $\mathcal{K}^n$ by

$$\eta(K) = \int_{\mathrm{Gr}(n,l)} \mu_k(K|V) \, \mathrm{d}\nu_l^n(V).$$

Evidently η is continuous, invariant, and homogeneous of degree k. It then follows from Corollary 9.1.2 that there exists $c \in \mathbf{R}$ such that $\eta = c\mu_k$. To compute the constant c we consider the case $K = B_n$:

$$c\mu_k(B_n) = \eta(B_n) = \int_{\mathrm{Gr}(n,l)} \mu_k(B_n|V) \, \mathrm{d}\nu_l^n(V) = \mu_k(B_l) \begin{bmatrix} n \\ l \end{bmatrix},$$

so that

$$
\begin{aligned}
c &= \frac{\mu_k(B_l)}{\mu_k(B_n)} \begin{bmatrix} n \\ l \end{bmatrix} \\[2mm]
&= \begin{bmatrix} l \\ k \end{bmatrix} \omega_k \begin{bmatrix} n \\ k \end{bmatrix}^{-1} \frac{1}{\omega_k} \begin{bmatrix} n \\ l \end{bmatrix} \\[2mm]
&= \frac{[l]!}{[k]![l-k]!} \frac{[k]![n-k]!}{[n]!} \frac{[n]!}{[l]![n-l]!} \\[2mm]
&= \frac{[n-k]!}{[n-l]![l-k]!} = \begin{bmatrix} n-k \\ l-k \end{bmatrix}.
\end{aligned}
$$

$\square$

Kubota's theorem can also be viewed combinatorially. For $0 \le k \le l$ and $K \in \mathcal{K}^l$, the mean projection formula (9.7) states that

$$\mu_k(K) = \int_{\mathrm{Gr}(l,k)} \mu_k(K|W) \, \mathrm{d}\nu_k^l(W).$$

Consequently,

$$\int_{\mathrm{Gr}(n,l)} \mu_k(K|V) \, \mathrm{d}\nu_l^n(V) = \int_{\mathrm{Gr}(n,l)} \int_{\mathrm{Gr}(V,k)} \mu_k((K|V)|W) \, \mathrm{d}\nu_k^l(W) \, \mathrm{d}\nu_l^n(V).$$

Since each $W \subseteq V$, we have $(K|V)|W = K|W$, so that

$$\int_{\mathrm{Gr}(n,l)} \mu_k(K|V) \, \mathrm{d}\nu_l^n(V) = \int_{\mathrm{Gr}(n,l)} \int_{\mathrm{Gr}(V,k)} \mu_k(K|W) \, \mathrm{d}\nu_k^l(W) \, \mathrm{d}\nu_l^n(V).$$

(9.8)

How does this last integral in (9.8) compare with the ordinary mean projection formula (9.7)? The difference is that in (9.8) we are 'counting' a k-subspace $W \in \mathrm{Gr}(n,k)$ once for each l-subspace $V \in \mathrm{Gr}(n,l)$

containing W. Ideally, our combinatorial intuition would suggest that counting l-subspaces that contain $W \in \text{Gr}(n, k)$ is the same as counting the $(l - k)$-subspaces of the quotient space $\mathbf{R}^n/W \cong \mathbf{R}^{n-k}$, of which there are

$$\begin{bmatrix} n - k \\ l - k \end{bmatrix}.$$

From this observation it would then follow that

$$\int_{\text{Gr}(n,l)} \int_{\text{Gr}(V,k)} \mu_k(K|W) \, d\nu_k^l(W) \, d\nu_l^n(V)$$

$$= \begin{bmatrix} n - k \\ l - k \end{bmatrix} \int_{\text{Gr}(n,k)} \mu_k(K|W) \, d\nu_k^n(W),$$

which is precisely what Kubota's theorem 9.4.3 asserts.

9.5 Mean cross-sectional volume

A k-dimensional *cross-section* of a compact convex set K is a subset of the form $K \cap V$, where $V \in \text{Graff}(K; k)$. Let $Y_k(K)$ denote the k-volume of a randomly chosen k-dimensional cross-section of K. The mean k-cross-sectional volume of K is given by the expectation $E(Y_k(K))$, which is computed by averaging over all $V \in \text{Graff}(K; k)$. In other words,

$$E(Y_k(K)) = \frac{1}{\lambda_k^n(\text{Graff}(K; k))} \int_{\text{Graff}(n,k)} \mu_k(K \cap V) \, d\lambda_k^n(V).$$

By Theorem 9.3.1 we have $\lambda_k^n(\text{Graff}(K; k)) = \mu_{n-k}(K)$, so that

$$E(Y_k(K))\mu_{n-k}(K) = \int_{\text{Graff}(n,k)} \mu_k(K \cap V) \, d\lambda_k^n(V)$$

$$= \int_{\text{Gr}(n,k)} \int_{V^\perp} \mu_k(K \cap (V + x)) \, dx \, d\nu_k^n(V).$$

Recall from elementary calculus that

$$\int_{V^\perp} \mu_k(K \cap (V + x)) \, dx = \mu_n(K),$$

for each $V \in \text{Gr}(n, k)$. It follows that

$$E(Y_k(K))\mu_{n-k}(K) = \int_{\text{Gr}(n,k)} \mu_n(K) \, d\nu_k^n(V)$$

$$= \mu_n(K) \begin{bmatrix} n \\ k \end{bmatrix}.$$

The mean k-cross-sectional volume of K is then given by

$$E(Y_k(K)) = \frac{\mu_n(K)}{\mu_{n-k}(K)} \begin{bmatrix} n \\ k \end{bmatrix}.$$

A one-dimensional cross-section is sometimes called a *chord*. Recall that $\mu_{n-1}(K) = (1/2)S(K)$, where $S(K)$ is the surface area of K. The expected length of a random chord in K is now given by

$$E(Y_1(K)) = \frac{\mu_n(K)}{\mu_{n-1}(K)} \begin{bmatrix} n \\ 1 \end{bmatrix} = \frac{2[n]\mu_n(K)}{S(K)}.$$

In particular, the expected length a random chord in the unit n-ball B is

$$E(Y_1(B)) = \frac{2[n]\omega_n}{n\omega_n} = \frac{\omega_n}{\omega_{n-1}}.$$

9.6 The Buffon needle problem revisited

We are now ready to generalize the Buffon needle problem to n-dimensional spaces. Suppose that $V \in \mathrm{Gr}(n, k)$ and that $\{v_1, \ldots, v_{n-k}\}$ is a basis for $V^\perp$. Let $\mathcal{V}$ denote the collection

$$\mathcal{V} = \{V + a_1 v_1 + \cdots + a_{n-k} v_{n-k} \ : \ a_1, \ldots, a_{n-k} \in \mathbf{Z}\}$$

of k-planes in $\mathbf{R}^n$. If a compact convex set K is moved randomly in $\mathbf{R}^n$, what is the expected number of intersections of K with the collection $\mathcal{V}$? In other words, if we denote by X_K the number of connected components of $gK \cap \mathcal{V}$ for a random Euclidean motion g, what is $E(X_K)$?

Let $\mathcal{L}$ denote the lattice in $V^\perp$ given by

$$\mathcal{L} = \{a_1 v_1 + \cdots + a_{n-k} v_{n-k} \ : \ a_1, \ldots, a_{n-k} \in \mathbf{Z}\},$$

and let C denote a fundamental domain of $\mathcal{L}$. Once again the concept of random motion is meaningful, since the symmetry of the collection $\mathcal{V}$ allows that we consider only motions involving translations by vectors in the fundamental domain C. Therefore, the collection of motions under consideration forms a compact Lie group, on which there exists a unique Haar probability measure.

Consider the set $\phi K + x$, where ϕ is a rotation and x is a vector. For a fixed $\phi \in O(n)$, the number of intersections of $\phi K + x$ with $\mathcal{V}$ is equal to the number of elements of $(\phi K + x)|V^\perp \cap \mathcal{L}$. Hence, the expected number of intersections of $\phi K + x$ with $\mathcal{V}$ is equal to

$$\frac{\mu_{n-k}((\phi K + x)|V^\perp)}{\mu_{n-k}(C)} = \frac{\mu_{n-k}((\phi K)|V^\perp)}{\mu_{n-k}(C)}, \tag{9.9}$$

by Theorem 8.5.2. Therefore, the expected number of intersections of $\phi K + x$ with $\mathcal{V}$ over all x and all ϕ is equal to the expected value of (9.9) over all orthogonal transformations ϕ. We now apply Corollary 9.4.2 to obtain

$$E(X_K) = \begin{bmatrix} n \\ n-k \end{bmatrix}^{-1} \frac{\mu_{n-k}(K)}{\mu_{n-k}(C)}.$$

As an example, consider the case in which K is a needle of length L in $\mathbf{R}^2$ and $\mathcal{V}$ is a collection of lines evenly spaced by a distance d. In this case $\mathcal{L}$ is a one-dimensional lattice of points on the line $V^{\perp}$, evenly spaced by a distance d. Since

$$\begin{bmatrix} 2 \\ 1 \end{bmatrix}^{-1} = \binom{2}{1}^{-1} \frac{\omega_1 \omega_1}{\omega_2} = \frac{2}{\pi},$$

it follows that

$$E(X_K) = \frac{2}{\pi} \frac{\mu_1(\text{needle})}{d} = \frac{2L}{\pi d},$$

which agrees with Buffon's original solution.

9.7 Intrinsic volumes on products

Next we use Hadwiger's characterization Theorem 9.1.1 to examine how the intrinsic volumes μ_i evaluate on orthogonal Cartesian products. We have the following generalization of Proposition 4.2.3.

Theorem 9.7.1 *Let $0 \leq k \leq n$, and suppose that $K \subseteq \mathbf{R}^k$ and $L \subseteq \mathbf{R}^{n-k}$ are polyconvex sets. Then*

$$\mu_i(K \times L) = \sum_{r+s=i} \mu_r(K)\mu_s(L). \tag{9.10}$$

Proof Evidently the set function $\mu_i(K \times L)$ is a continuous valuation in each of the variables K and L when the other is held fixed. Note that each motion $\phi \in E_k$ of $\mathbf{R}^k$ is the restriction of a motion $\Phi \in E_n$ that restricts to the identity on the complementary space $\mathbf{R}^{n-k}$. Therefore,

$$\mu_i(\phi K \times L) = \mu_i(\Phi(K \times L)) = \mu_i(K \times L).$$

In other words, the set function $\mu_i(K \times L)$ is a continuous invariant valuation in each variable. By an iteration of Theorem 9.1.1 in each variable, there exist constants $c_{rs} \in \mathbf{R}$ such that

$$\mu_i(K \times L) = \sum_{r=0}^{k} \sum_{s=0}^{n-k} c_{rs} \mu_r(K) \mu_s(L)$$

for all $K \in \mathcal{K}^k$ and $L \in \mathcal{K}^{n-k}$. Let C_m denote the unit m-dimensional cube. For $\alpha, \beta \geq 0$,

$$\begin{aligned}
\mu_i(\alpha C_k \times \beta C_{n-k}) &= \sum_{r=0}^{k} \sum_{s=0}^{n-k} c_{rs} \mu_r(C_k) \mu_s(C_{n-k}) \alpha^r \beta^s \\
&= \sum_{r=0}^{k} \sum_{s=0}^{n-k} c_{rs} \binom{k}{r} \binom{n-k}{s} \alpha^r \beta^s.
\end{aligned}$$

Meanwhile,

$$\mu_i(\alpha C_k \times \beta C_{n-k}) = \sum_{r+s=i} \binom{k}{r} \binom{n-k}{s} \alpha^r \beta^s,$$

by Proposition 4.2.3. Therefore, for $0 \leq r \leq k$ and $0 \leq s \leq n-k$, we have $c_{rs} = 1$ if $r + s = i$ and $c_{rs} = 0$ otherwise. Hence,

$$\mu_i(K \times L) = \sum_{r=0}^{k} \sum_{s=0}^{n-k} c_{rs} \mu_r(K) \mu_s(L) = \sum_{r+s=i} \mu_r(K) \mu_s(L).$$

$\square$

Corollary 9.7.2 *Suppose that μ is a convex-continuous invariant valuation on* Polycon(n) *such that*

$$\mu(K \times L) = \mu(K)\mu(L), \tag{9.11}$$

for all $K \subseteq \mathbf{R}^k$ and $L \subseteq \mathbf{R}^{n-k}$, where $0 \leq k \leq n$. Then either $\mu = 0$ or there exists $c \in \mathbf{R}$ such that

$$\mu = \mu_0 + c\mu_1 + c^2\mu_2 + \cdots + c^n\mu_n. \tag{9.12}$$

Conversely, if μ is a valuation of the form (9.12) then μ also satisfies the multiplicative rule (9.11).

Proof By Hadwiger's Theorem 9.1.1, there exist constants $c_i \in \mathbf{R}$ such that

$$\mu = c_0\mu_0 + c_1\mu_1 + c_2\mu_2 + \cdots + c_n\mu_n.$$

For $k \in \{0, 1, \ldots, n\}$ denote by C_k the unit k-cube in $\mathbf{R}^k$. Then, for all $\alpha, \beta \geq 0$, the condition (9.11) implies that

$$\mu(\alpha C_k \times \beta C_{n-k}) = \mu(\alpha C_k)\mu(\beta C_{n-k})$$

$$= \left(\sum_{r=0}^{k} c_r \alpha^r \mu_r(C_k)\right)\left(\sum_{s=0}^{n-k} c_s \beta^s \mu_s(C_{n-k})\right)$$

$$= \sum_{r=0}^{k}\sum_{s=0}^{n-k} c_r c_s \mu_r(C_k)\mu_s(C_{n-k})\alpha^r \beta^s.$$

Meanwhile, by Theorem 9.7.1,

$$\mu(\alpha C_k \times \beta C_{n-k}) = \sum_{i=0}^{n} c_i \mu_i(\alpha C_k \times \beta C_{n-k})$$

$$= \sum_{i=0}^{n} c_i \sum_{r+s=i} \mu_r(C_k)\mu_s(C_{n-k})\alpha^r \beta^s$$

$$= \sum_{r=0}^{k}\sum_{s=0}^{n-k} c_{r+s}\mu_r(C_k)\mu_s(C_{n-k})\alpha^r \beta^s.$$

Therefore, $c_{r+s} = c_r c_s$ for all $0 \leq r, s \leq n$. In particular, $c_0 = c_{0+0} = c_0^2$, so that $c_0 = 0$ or $c_0 = 1$. If $c_0 = 0$ then $c_r = c_{r+0} = c_r c_0 = 0$, so that $\mu = 0$. If $c_0 = 1$ then relabel $c = c_1$. For $r > 0$, we then have $c_r = c_{1+1+\cdots+1} = c_1^r = c^r$, from which (9.12) follows. The converse follows from Theorem 9.7.1 by means of a similar argument. $\qquad\square$

We conclude this chapter with another example of a convex-continuous invariant valuation. Choose a real-valued non-negative continuous function $f(t)$ of a non-negative variable t, such that $f(t)$ is decreasing sufficiently fast to 0 as $t \to \infty$.

For any n, and for any non-empty compact convex set K in $\mathbf{R}^n$, set

$$\mu(f; K) = \int_{\mathbf{R}^n} f(d(p, K))\, \mathrm{d}p$$

where $d(p, K)$ is the distance from the point p to the set K, and where $\mathrm{d}p$ denotes the ordinary n-dimensional Lebesgue measure. We show that μ satisfies the inclusion–exclusion principle; that is,

$$\mu(f; K_1 \cup \cdots \cup K_m) = \sum_i \mu(f; K_i) - \sum_{i<j} \mu(f; K_i \cap K_j) + \cdots \quad (9.13)$$

whenever $K_1, K_2, \ldots, K_m$ and $K_1 \cup \cdots \cup K_m$ are convex. We begin with a direct proof of the case $m = 2$.

Let $p \in \mathbf{R}^n$. Since K_1 and K_2 are compact, there exist unique points $q_1 \in K_1$ and $q_2 \in K_2$ such that $d(p, q_i) = d(p, K_i)$. We assume that $K_1 \cup K_2$ is convex, so that $K_1 \cap K_2 \neq \emptyset$. Evidently, $d(p, K_1 \cup K_2) = \min\{d(p, K_1), d(p, K_2)\}$, and $d(p, K_1 \cap K_2) \geq \max\{d(p, K_1), d(p, K_2)\}$. Since $K_1 \cup K_2$ is convex, the line segment I connecting q_1 and q_2 must intersect $K_1 \cap K_2$ at some point q'. Either $d(p, q') \leq d(p, q_1)$ or $d(p, q') \leq d(p, q_2)$. Therefore,

$$d(p, K_1 \cap K_2) \leq d(p, q') \leq \max\{d(p, K_1), d(p, K_2)\}$$

as well, so that $d(p, K_1 \cap K_2) = \max\{d(p, K_1), d(p, K_2)\}$. We then obtain

$$\begin{aligned}
f(d(p, K_1 \cup K_2)) &+ f(d(p, K_1 \cap K_2)) \\
&= f(\min\{d(p, K_1), d(p, K_2)\}) + f(\max\{d(p, K_1), d(p, K_2)\}) \\
&= f(d(p, K_1)) + f(d(p, K_2)),
\end{aligned}$$

for all $p \in \mathbf{R}^n$. Hence,

$$\mu(f; K_1 \cup K_2) = \mu(f; K_1) + \mu(f; K_2) - \mu(f; K_1 \cap K_2).$$

Since f is a continuous function, it follows easily from the definition of the Hausdorff topology that μ is a continuous function of K. The general case of (9.13) then follows from Theorems 5.1.1 and 2.2.1. Thus we can extend $\mu(f; K)$ to an invariant valuation defined on all polyconvex sets K. The expression of $\mu(f; K)$ as a linear combination of the intrinsic volumes (given by Theorem 9.1.1) exhibits the *moments*

$$m_j(f) = \int_0^\infty x^j f(x) \, \mathrm{d}x$$

of the function f and has the form

$$\mu(f; K) = f(0)\mu_n(K) + \sum_{i=1}^n c_i m_{i-1}(f)\mu_{n-i}(K),$$

where the constants c_i are independent of the function f.

To understand this, and to compute the c_i, consider the case of the n-dimensional ball αB of radius α, centered at the origin. From the definition, we have

$$\mu(f; \alpha B) \;=\; \int_{\mathbf{R}^n} f(d(p, \alpha B)) \, \mathrm{d}p$$

$$= \int_{\alpha B} f(0)\, dp + \int_{\mathbf{R}^n - \alpha B} f(|p| - \alpha)\, dp$$

$$= f(0)\mu_n(\alpha B) + \int_{u \in \mathbf{S}^{n-1}} \int_\alpha^\infty f(r - \alpha) r^{n-1}\, dr\, du$$

$$= f(0)\mu_n(\alpha B) + n\omega_n \int_\alpha^\infty f(r - \alpha) r^{n-1}\, dr$$

$$= f(0)\mu_n(\alpha B) + n\omega_n \int_0^\infty f(y)(y + \alpha)^{n-1}\, dy$$

$$= f(0)\mu_n(\alpha B) + n\omega_n \sum_{j=0}^{n-1} \binom{n-1}{j} \int_0^\infty f(y) y^j \alpha^{n-1-j}\, dy$$

$$= f(0)\mu_n(\alpha B) + n\omega_n \sum_{i=1}^{n} \binom{n-1}{i-1} \alpha^{n-i} m_{i-1}(f)$$

$$= f(0)\mu_n(\alpha B) + \sum_{i=1}^{n} \binom{n-1}{i-1} \frac{n\omega_n m_{i-1}(f)}{\mu_{n-i}(B)} \mu_{n-i}(\alpha B).$$

Once again ω_n denotes the volume of the unit ball B, so that the surface area of the unit sphere $\mathbf{S}^{n-1}$ is given by $n\omega_n$. It follows that

$$c_i = \binom{n-1}{i-1} \frac{n\omega_n}{\mu_{n-i}(B)} = \binom{n-1}{i-1} n\omega_n \binom{n}{n-i}^{-1} \frac{\omega_i}{\omega_n} = i\omega_i,$$

so that

$$\mu(f; K) = f(0)\mu_n(K) + \sum_{i=1}^{n} i\omega_i m_{i-1}(f)\mu_{n-i}(K). \tag{9.14}$$

A notable special case is obtained by setting $f(x) = e^{-\pi x^2}$. The resulting valuation is called the *Wills functional* and is denoted by $W(K)$.

Theorem 9.7.3 *If $K \subseteq V$ and $L \subseteq V^\perp$ for some linear variety V, then*

$$W(K \times L) = W(K)W(L).$$

Proof The moment m_{i-1} of $f(x) = e^{-\pi x^2}$ is given by

$$m_{i-1} = \int_0^\infty x^{i-1} e^{-\pi x^2}\, dx.$$

On substituting $u = \pi x^2$, we have

$$m_{i-1} = \frac{1}{2\pi^{i/2}} \int_0^\infty u^{\frac{i}{2}-1} e^{-u} \, du = \frac{1}{2\pi^{i/2}} \Gamma\left(\frac{i}{2}\right) = \frac{1}{i\omega_i}.$$

On substituting into (9.14), we have

$$\mu(f; K) = \mu_n(K) + \sum_{i=1}^n i\omega_i m_{i-1}(f) \mu_{n-i}(K) = \sum_{i=0}^n \mu_{n-i}(K). \quad (9.15)$$

It then follows from Corollary 9.7.2 that $W(K \times L) = W(K)W(L)$.
$\square$

From Theorem 9.7.3 we obtain the following characterization for the Wills functional W. Once again let C_n denote the unit n-cube in $\mathbf{R}^n$.

Corollary 9.7.4 *Suppose that μ is a convex-continuous invariant valuation on* Polycon(n) *such that*

$$\mu(K \times L) = \mu(K)\mu(L),$$

for all $K \subseteq \mathbf{R}^k$ and $L \subseteq \mathbf{R}^{n-k}$, where $0 \leq k \leq n$. If $\mu(C_1) = 2$ then $\mu(K) = W(K)$ for all $K \in$ Polycon(n).

Proof Since $\mu(C_1) \neq 0$, Corollary 9.7.2 implies the existence of some $c \in \mathbf{R}$ such that

$$\mu = \mu_0 + c\mu_1 + c^2\mu_2 + \cdots + c^n\mu_n = \sum_{i=0}^n c^i \mu_i.$$

Hence,

$$2 = \mu(C_1) = \sum_{i=0}^n c^i \mu_i(C_1) = \mu_0(C_1) + c\mu_1(C_1) = 1 + c,$$

so that $c = 1$. It then follows from (9.15) that $\mu = W$. $\square$

9.8 Computing the intrinsic volumes

The actual computation of the intrinsic volumes of a polyconvex set is difficult in general. It is difficult even for (ordinary) volume, and also for the surface area, which is given by $2\mu_{n-1}$. It can be difficult even in low dimensions: consider the perimeter of an ellipse, for example.

However, we are already able to compute the intrinsic volumes of a large class of compact convex sets. Theorem 4.2.1 gives a formula for the kth intrinsic volume of an orthogonal parallelotope in $\mathbf{R}^n$, and

Theorem 9.2.4 gives the intrinsic volumes of the unit ball. Theorem 9.7.1 gives a formula for computing the intrinsic volume of an orthogonal Cartesian product of two polyconvex sets whose intrinsic volumes are already known.

In this section we work with two additional classes of polyconvex sets. First, we compute the intrinsic volumes of an arbitrary (possibly non-orthogonal) parallelotope. We then comment on the intrinsic volumes of an arbitrary convex polytope and give an explicit formula for the first intrinsic volume μ_1 of a convex polytope in $\mathbf{R}^3$ in terms of the lengths of its edges and the angles between the outer normals to its adjacent facets.

In Section 4.2 we proved Theorem 4.2.4, which characterized volume on $\mathrm{Par}(n)$. Recall that $\mathrm{Par}(n)$ denoted the lattice of finite unions of orthogonal parallelotopes having edges parallel to the coordinate axes of $\mathbf{R}^n$. A similar theorem holds for more general parallelotopes.

Let $v_1, \ldots, v_n$ be a basis for $\mathbf{R}^n$, and let $\mathrm{Par}(v_1, \ldots, v_n)$ denote the lattice of finite unions of parallelotopes with edges parallel to the vectors v_i. The characterization theorem for volume in $\mathrm{Par}(n)$, Theorem 4.2.4, generalizes immediately to the lattice $\mathrm{Par}(v_1, \ldots, v_n)$.

Theorem 9.8.1 (The volume theorem for $\mathrm{Par}(v_1, \ldots, v_n)$) *Let μ be a translation invariant simple valuation defined on $\mathrm{Par}(v_1, \ldots, v_n)$, and suppose that μ is either continuous or monotone. Then there exists $c \in \mathbf{R}$ such that $\mu(P) = c\mu_n(P)$ for all $P \in \mathrm{Par}(v_1, \ldots, v_n)$; that is, μ is equal to the volume, up to a constant factor.*

Proof For each i, let $\overline{v}_i$ denote the line segment having endpoints at v_i and the origin o. The proof of Theorem 9.8.1 is the same as the proof of Theorem 4.2.4, provided that we replace the unit cube $[0,1]^n$ with the 'unit parallelotope'

$$C = \overline{v}_1 + \cdots + \overline{v}_n$$

of $\mathrm{Par}(v_1, \ldots, v_n)$. It then follows that $\mu = c\mu_n$, where $c = \mu(C)/\mu_n(C)$.
$\square$

Theorem 9.8.1 leads in turn to a formula for the kth intrinsic volume of an arbitrary parallelotope. Note that if $P \in \mathrm{Par}(v_1, \ldots, v_n)$ is a parallelotope, then P is a translate of a parallelotope of the form

$$a_1\overline{v}_1 + \cdots + a_n\overline{v}_n,$$

for some $a_1, \ldots, a_n \geq 0$.

Theorem 9.8.2 *For all $1 \leq k \leq n$ and all $a_1, \ldots, a_n \geq 0$,*

$$\mu_k(a_1 \overline{v}_1 + \cdots + a_n \overline{v}_n) = \sum_{1 \leq i_1 < \cdots < i_k \leq n} \mu_k(a_{i_1} \overline{v}_{i_1} + \cdots + a_{i_k} \overline{v}_{i_k}).$$

The value of each term $\mu_k(a_{i_1} \overline{v}_{i_1} + \cdots + a_{i_k} \overline{v}_{i_k})$ is easily computed using elementary linear algebra. If A is a $k \times n$ matrix whose jth row is given by the coordinates of the vector $a_{i_j} v_{i_j}$, for $j = 1, \ldots, k$, then

$$\mu_k(a_{i_1} \overline{v}_{i_1} + \cdots + a_{i_k} \overline{v}_{i_k}) = \sqrt{\det{(AA^T)}}.$$

See, for example, [94, p. 234].
Proof of Theorem 9.8.2 Define

$$\nu_k(P) = \sum_{1 \leq i_1 < \cdots < i_k \leq n} \mu_k(a_{i_1} \overline{v}_{i_1} + \cdots + a_{i_k} \overline{v}_{i_k}),$$

and extend ν_k to a valuation on all of $\mathrm{Par}(v_1, \ldots, v_n)$ with Theorem 4.1.3. Let $\eta = \mu_k - \nu_k$. We will prove that $\eta(P) = 0$, for all $P \in \mathrm{Par}(v_1, \ldots, v_n)$.

Since both μ_k and ν_k vanish in dimensions less than k, it follows that η vanishes in dimensions less than k as well. If we restrict η to the k-plane $V_{i_1, \ldots, i_k}$ spanned by the (independent) vectors $v_{i_1}, \ldots, v_{i_k}$, then η becomes a continuous translation invariant *simple* valuation on $\mathrm{Par}(v_{i_1}, \ldots, v_{i_k})$. It follows from Theorem 9.8.1 (in dimension k) that there exists $c \in \mathbf{R}$ such that $\eta(P) = c\mu_k(P)$ for all $P \in \mathrm{Par}(v_{i_1}, \ldots, v_{i_k})$. However,

$$\eta(\overline{v}_{i_1} + \cdots + \overline{v}_{i_k}) = 0$$

by the definitions of η and ν_k. Therefore $c = 0$, and η vanishes in dimension k.

In other words, if we restrict η to a $k + 1$-plane $V_{i_1, \ldots, i_{k+1}}$ spanned by the vectors $v_{i_1}, \ldots, v_{i_{k+1}}$, then we have once again a continuous translation invariant simple valuation, this time on a lattice of parallelotopes in a space of dimension $k + 1$. Theorem 9.8.1 again applies, and there exists $c \in \mathbf{R}$ such that $\eta(P) = c\mu_{k+1}(P)$ for all $P \in \mathrm{Par}(v_{i_1}, \ldots, v_{i_{k+1}})$. However, η is homogeneous of degree k, whereas μ_{k+1} is homogeneous of degree $k + 1$. Hence, we have $c = 0$, and η vanishes in dimension $k + 1$. Continuing this argument in each higher dimension we conclude that $\eta(P) = 0$ for P of any dimension in $\mathrm{Par}(v_1, \ldots, v_n)$. $\square$

Computing the intrinsic volumes of arbitrary polytopes is more difficult. Recall from Steiner's formula 9.2.3 that, for all $r \geq 0$,

$$\mu_n(P + rB) = \sum_{i=0}^{n} \mu_i(P)\omega_{n-i}r^{n-i}, \qquad (9.16)$$

where B denotes the unit ball in $\mathbf{R}^n$. Let us consider the convex set $P + rB$ more carefully. If $x \in P + rB$, then there exists a unique point $x_P \in P$ such that

$$|x - x_P| \leq |x - y|,$$

for all $y \in P$. This follows from the fact that P is compact and convex. If $x \in P$, then evidently $x = x_P$. If $x \notin P$, then x_P lies on the boundary ∂P of P. Moreover, if $x \notin P$ and $y \in \partial P$, then $y = x_P$ if and only if $x - y \perp H$, where H is a support plane of P and $y \in P \cap H$.

Let $P_i(r)$ denote the set of all $x \in P + rB$ such that x_P lies in the relative interior of an i-face of P. So, for example, $P_n(r) = \text{int}(P)$. Evidently

$$P + rB = \bigcup_{i=0}^{n} P_i(r), \qquad (9.17)$$

a disjoint union.

Denote by $F_i(P)$ the set of all i-dimensional faces of P. For each proper face Q of P denote by $M(Q, r)$ the set

$$M(Q, r) = \{y + \delta v \ : \ 0 \leq \delta \leq r\},$$

where $y \in \text{relint}(Q)$ and v is any outward unit normal to ∂P at the point y. See Figure 9.1.

It turns out that, for $0 \leq i < n$,

$$P_i(r) = \bigcup_{Q \in F_i(P)} M(Q, r).$$

To see this, suppose that $x \in P_i(r)$. Then $x_P \in \text{relint}(Q)$ for some i-face Q of P. If $x \in \text{relint}(Q)$, then $x \in M(Q, r)$. If $x \notin \text{relint}(Q)$, then $x \neq x_P$. Let $v = (x - x_P)/|x - x_P|$. Then $v \perp H$ for some support plane H of P at x_P, and $x = x_P + \delta v$ for $0 < \delta = |x - x_P| \leq r$. In other words, $x \in M(Q, r)$.

Conversely, suppose that $x = y + \delta v \in M(Q, r)$ for some i-face Q. If $\delta = 0$ then $x \in \text{relint}(Q)$. Otherwise, we have $x - y \perp H$ for some support plane H at the point $y \in \text{relint}(Q)$. It follows that $y = x_P$ and $x \in P_i(r)$.

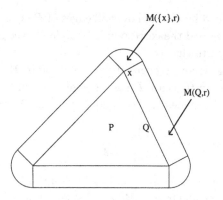

Fig. 9.1. The Decomposition of $P + rB$.

Given a subset A of $\mathbf{R}^n$, the *affine hull* of A is the intersection of all planes in $\mathbf{R}^n$ (not necessarily through the origin) containing A. Let $A^\perp$ denote the set of vectors in $\mathbf{R}^n$ orthogonal to the affine hull of A.

If Q is an i-face then $Q^\perp$ has dimension $n - i$, so that $\mu_n(M(Q, r)) = r^{n-i}\mu_n(M(Q, 1))$ It follows that $\mu_n(P_i(r)) = r^{n-i}\mu_n(P_i(1))$. From (9.17) we then obtain

$$\mu_n(P + rB) = \sum_{i=0}^{n} \mu_n(P_i(1))r^{n-i}, \tag{9.18}$$

for all $r > 0$. By comparing (9.16) and (9.18) we find that

$$\mu_i(P) = \frac{\mu_n(P_i(1))}{\omega_{n-i}}. \tag{9.19}$$

For example, consider a convex polyhedron P in $\mathbf{R}^3$ having edges $\bar{z}_1, \ldots, \bar{z}_m$. Each edge $\bar{z}_i$ is formed by the meeting of two facets Q_{i_1} and Q_{i_2} of P, having outward unit normals u_{i_1} and u_{i_2} respectively. Therefore, the volume of $M(\bar{z}_i, 1)$ is given by

$$\mu_3(M(\bar{z}_i, 1)) = \mu_1(\bar{z}_i)\frac{\arccos(u_{i_1} \cdot u_{i_2})}{2} = \frac{\mu_1(\bar{z}_i)\theta_i}{2},$$

where θ_i is the angle between the outer normals to the facets Q_{i_1} and Q_{i_2}. Since $\omega_{3-1} = \omega_2 = \pi$, it follows that

$$\mu_1(P) = \frac{1}{2\pi}\sum_{i=1}^{m} \mu_1(\bar{z}_i)\theta_i.$$

In other words, the first intrinsic volume of a polyhedron P in $\mathbf{R}^3$ is given by $1/(2\pi)$ multiplied by the sum over all edges of P of the product of the length of each edge and the corresponding angle between the normals to the facets adjacent to that edge.

Suppose that P is an orthogonal parallelotope in $\mathbf{R}^3$ having dimensions $a_1 \times a_2 \times a_3$. In this case P has 12 edges, four of length a_i for each $i = 1, 2, 3$, while the angles θ_i are all equal to $\pi/2$. We then obtain

$$\mu_1(P) = \frac{1}{2\pi} \sum_{i=1}^{3} 4a_i \frac{\pi}{2} = a_1 + a_2 + a_3,$$

as asserted by Theorem 4.2.1. Similarly, it is not difficult to show that if T_s is a regular tetrahedron in $\mathbf{R}^3$ of edge length s, then

$$\mu_1(T_s) = \frac{3s}{\pi} \arccos\left(-\frac{1}{3}\right).$$

formulas for μ_1 (under a slightly different normalization) evaluated on a variety of common solids in $\mathbf{R}^3$ are described by Hadwiger in [38, pp. 36–37].

9.9 Notes

For the original proof of Hadwiger's characterization Theorem 9.1.1, see [39]. In addition to providing straightforward verification of many results in integral geometry, Theorem 9.1.1 also provides a connection between rigid motion invariant set functions and symmetric polynomials [13].

The normalization λ_k^n of the invariant measures on $\mathrm{Graff}(n, k)$ which we have used in the text is natural for $\mathbf{R}^\infty$; it leads to a good definition of the intrinsic volumes in Hilbert space. The invariant measure under $O(n)$ in $\mathrm{Gr}(n, k)$, corresponding to the measure λ_k^n on $\mathrm{Graff}(n, k)$ under E_n, is the Gauss–Wiener measure (as can be easily verified).

The article by Schanuel [83] provides an interesting and accessible perspective on Steiner's formula and the first intrinsic volume, or generalized length, μ_1, for compact polyhedra in $\mathbf{R}^3$. See also [38, pp. 36–37] for extensive computations of μ_1 in $\mathbf{R}^3$.

Steiner's formula (Theorem 9.2.3) is a special case of the polynomial formula for *mixed volumes*. For $K_1, K_2, \ldots, K_m \in \mathcal{K}^n$, and real numbers $\lambda_1, \lambda_2, \ldots, \lambda_m > 0$, the *Minkowski linear combination* $K = \lambda_1 K_1 + \lambda_2 K_2 + \cdots + \lambda_m K_m$ is the convex body consisting of the set

$$\{\lambda_1 x_1 + \lambda_2 x_2 + \cdots + \lambda_m x_m : x_i \in K_i\}.$$

The basis of the theory of mixed volumes is the polylinearization of volume with respect to Minkowski linear combinations. If $K_1, \ldots, K_m \in \mathcal{K}^n$ and $\lambda_1, \ldots, \lambda_m > 0$, then the volume μ_n is a homogeneous polynomial in the positive variables $\lambda_1, \ldots, \lambda_m$; that is,

$$\mu_n(\lambda_1 K_1 + \cdots + \lambda_m K_m) = \sum_{i_1, \ldots, i_n = 1}^{m} V(K_{i_1}, \ldots, K_{i_n}) \lambda_{i_1} \cdots \lambda_{i_n}, \quad (9.20)$$

where each symmetric coefficient $V(K_{i_1}, \ldots, K_{i_n})$ depends only on the bodies $K_{i_1}, \ldots, K_{i_n}$.

Given $K_1, \ldots, K_n \in \mathcal{K}^n$, the coefficient $V(K_1, \ldots, K_n)$ is called the *mixed volume* of the convex bodies $K_1, \ldots, K_n$. It is well known, but not trivial, that the mixed volume $V(K_1, \ldots, K_n)$ is a non-negative continuous function in n variables on the set $\mathcal{K}^n$, symmetric in the variables K_i, and monotonic with respect to the subset partial ordering on $\mathcal{K}^n$. Moreover, the mixed volume $V(K_1, \ldots, K_n)$ is a valuation in each of its parameters K_i, provided that the other parameters are held fixed. The mixed volumes also satisfy several useful inequalities, such as the Brunn–Minkowski and Aleksandrov–Fenchel inequalities. The intrinsic volumes μ_i studied in the present work are normalizations of the mixed volume of a convex body K with the unit ball B_n. Specifically,

$$\mu_i(K) = \binom{n}{i} \frac{1}{\omega_{n-i}} V(\underbrace{K, \ldots, K}_{i}, \underbrace{B_n, \ldots, B_n}_{n-i}),$$

for all $K \in \mathcal{K}^n$. For a thorough treatment of the theory of mixed volumes, see [85].

Theorem 9.3.2 is originally due to Crofton [14, 15, 16] (see also [85, p. 235]). Crofton's formula generalizes the ideas of Barbier presented in Chapter 1. Theorem 9.4.3 was proved by Kubota in [58]; see also [85, p. 295]. Numerous variations of the formulas of Crofton and Kubota can also be found in [1, 2, 82, 99].

Wills first defined the functional $W(K)$ in the context of lattice point enumeration [101]. For a discussion of the Wills functional and its applications, see also [85, p. 302] and [96].

There exist many important functionals on $\mathcal{K}^n$ that are invariant with respect to rigid motions, but that elude Hadwiger's characterization. One striking example, pointed out by Hadwiger in [38, p. 44], is the *affine surface area*, denoted Ω. Although originally defined only for polytopes and smooth convex bodies [85, p. 419], the affine surface area was later extended to an affine invariant functional on all of $\mathcal{K}^n$ (see, for example, [20, 59, 60, 61, 66, 89, 100]). Extended affine surface area was shown to be a valuation by Schütt [88]. It turns out, however, that affine surface area cannot be expressed as a linear combination of intrinsic volumes, the reason being that it is not continuous, but only upper semi-continuous, on $\mathcal{K}^n$. Instead, the affine surface area satisfies a very strong invariance property: If $K \in \mathcal{K}^n$, then $\Omega(\phi K) = \Omega(K)$, for all affine transformations ϕ; that is, for all translations and linear transformations having unit determinant. The affine surface area also satisfies a number of useful inequalities [67]. The question of how to characterize the affine surface area as a valuation possessing certain properties (in the style of Hadwiger's characterization theorem) remains open.

Another interesting open problem of convex set theory is the problem of classification of invariant set functions that are convex-continuous and of *polynomial type*. For example, a *bivaluation* $\mu(K, L)$, defined for polyconvex sets K and L, is a real-valued set function that is a valuation in either variable, provided that the other variable is held fixed. In a similar vein one defines a trivaluation and more generally an n-valuation.

If μ is an n-valuation, set (for K convex)

$$\nu(K) = \mu(K, K, \ldots, K).$$

The set function ν is called a homogeneous polynomial set function of degree n. A finite sum of homogeneous polynomial set functions is called a *set function of polynomial type*.

It would be interesting to give an 'intrinsic' definition of set functions of polynomial type. It is tempting to conjecture that the classification of invariant set functions of polynomial type is related to symmetric functions. For example, if $f(x_1, \ldots, x_n)$ is any homogeneous symmetric polynomial, and if one sets $\nu(P) = f(x_1, \ldots, x_n)$ for any parallelotope with side lengths $x_1, \ldots, x_n$, one can extend ν to all polyconvex sets by using the expression of f in terms of elementary symmetric functions and then replacing each occurrence of an elementary symmetric function by the corresponding intrinsic volume. Does one get all invariant set functions of polynomial type in this way? (See [13].)

Hadwiger's characterization Theorem 9.1.1 and Theorem 9.3.1 together suggest yet another striking interpretation in the context of simplicial complexes. Recall that elements of the distributive lattice $L(S)$ of Chapter 3 consist of finite unions of *simplices*, a simplex being a collection of subsets of a finite set S which is closed under the taking of subsets; i.e. an *order ideal* with respect to subset inclusion.

The distributive lattice Polycon(n) admits a partial analogy to this notion. Loosely speaking, we can think of Polycon(n) as a collection of simplicial complexes, such that the simplices are the compact convex sets in $\mathbf{R}^n$, and the faces of such a simplex (a convex set) are its *intersections* with lower dimensional planes; i.e. the k-faces of an n-dimensional body K would be the sets $K \cap V$ for $V \in \text{Graff}(n, n-k)$. Note, however, the codimensional duality.

More precisely, define a *filter* F of a partially ordered set P to be a subset of P such that if $x \in F$ and $y \geq x$ then $y \in F$. In other words, a filter of P is an order ideal of the dual P^* of P (in which all of the order relations in P are reversed). In dual analogy to order ideals, the set of *minimal* elements of a filter F is an antichain in P, and we can view the filters of P as simplicial complexes, such that a simplex is now defined to be a filter with a unique minimal element. As with order ideals, the collection of all filters in P forms a distributive lattice.

Now consider the case $P = \text{Aff}(n)$, ordered by subset inclusion. For $V \in \text{Aff}(n)$, denote by $\check{V}$ the filter consisting of all linear varieties in $\mathbf{R}^n$ containing V; that is, $\check{V} = \{W \in \text{Aff}(n) : V \subseteq W\}$. One may view $\check{V}$ as a simplex in the distributive lattice of filters in $\text{Aff}(n)$. Note that, although this lattice is also graded from 0 to n, this grading is dual to that of $\text{Aff}(n)$. In other words, the k-faces of $\check{V}$ are the $(n-k)$-dimensional linear varieties in $\check{V}$.

If $K \in \text{Polycon}(n)$ and $V \cap K \neq \emptyset$, then $W \cap K \neq \emptyset$ for all $W \in \check{V}$. This motivates us to define $\check{K} = \{V \in \text{Aff}(n) : V \cap K \neq \emptyset\}$. Evidently, $\check{K}$ is a filter in $\text{Aff}(n)$. Consequently one may view $\check{K}$ as a simplicial complex in the lattice of filters of $\text{Aff}(n)$, and it is then natural to ask how the lattice $\text{Polycon}(n)$ relates to the lattice of filters of $\text{Aff}(n)$. (Is it a quotient lattice?)

Viewed in this light, the intrinsic volumes μ_k become the polyconvex analogues of the face enumerators (also denoted μ_k) of Section 3.2. This analogy is buttressed by the formula (9.4), which states that the intrinsic volume μ_k is, in some sense, an enumerator (measure) of the 'k-faces' of a 'simplex' (convex set) K in $\text{Polycon}(n)$.

In a similar spirit, one may view the discrete basis Theorem 3.2.4 and Hadwiger's characterization Theorem 9.1.1 as entirely analogous classifications of invariant valuations on the discrete and polyconvex lattices, respectively. The analogy is pressed further by the existence of Helly-type theorems (Theorem 3.3.1 and Theorem 5.3.3) in both contexts. Such remarkable similarities suggest the possibility of additional ties between the simplicial and order structures of these two lattices. For example, do there exist polyconvex analogues of the binomial coefficients, and if so, how are these analogues related to the flag coefficients of Chapter 6?

Similar observations may be made concerning the lattice $\text{Mod}(n)$. Although the discrete basis Theorem 3.2.4, Theorem 4.2.5 and Hadwiger's characterization Theorem 9.1.1 characterize all suitably invariant valuations on the lattices $L(S)$, $\text{Par}(n)$, and $\text{Polycon}(n)$ respectively, there remains the question of how to characterize invariant valuations on simplices in $\text{Mod}(n)$, as defined in Section 6.4. Indeed the success of the lattice analogy for $\text{Mod}(n)$ in Chapter 6 suggests the possibility of a comprehensive theory of simplicial complexes in $\text{Mod}(n)$ that has yet to be developed.

The extension of the intrinsic volumes from elementary symmetric functions on parallelotope edge lengths (see Theorem 4.2.1) to invariant valuations on compact convex sets motivates the following question regarding total positivity. Let A be an $n \times n$ matrix of real numbers, and denote by A_{ij} the entry of A in the ith row and jth column. The matrix A is said to be *totally positive* if $\det \widetilde{A} \geq 0$ for all minors $\widetilde{A}$ of A. See, for example, [47].

For $K \in \mathcal{K}^n$, denote by $A(K)$ the matrix with entries $A_{ij}(K) = \mu_{j-i}(K)$ if $j - i \geq 0$; otherwise let $A_{ij}(K) = 0$. An interesting open question is that of whether the matrix $A(K)$ is totally positive. For example, if we let K denote an orthogonal parallelotope with dimensions $a_1 \times \cdots \times a_n$, then the total positivity condition on the 2×2 minors of $A(K)$ is equivalent to Newton's inequality (see [69, p. 106] or [41, pp. 51–55]) for the elementary symmetric functions $e_1, \ldots, e_n$ of the variables a_i:

$$e_i^2 \geq e_{i+1} e_{i-1}.$$

For general compact convex sets K, it may be necessary to replace the intrinsic volumes by a renormalization of the μ_i in order to obtain total positivity for the matrix $A(K)$.

Among related open questions, the problem of syzygies seems particularly difficult. Let $A_1, \ldots, A_j$ be compact convex sets, and let $I_{A_i}^{(k)}$ be the indicator functions of the set $\text{Graff}(A; k)$ on $\text{Graff}(n, k)$. The problem is to find a basis for all linear relations

$$\sum_i \alpha_i I_{A_i}^{(k)} = 0$$

for all $k \geq 0$ (so that $I_A^0 = I_A$). Even for $k = 0$, where $I_A^{(k)} = I_A$, the problem is not entirely trivial (though it has been solved fully). The restricted inclusion–exclusion principle for convex sets yields some such identities, but there are probably some more special identities for each k.

No doubt there exist additional lattices, for which these questions may be asked and answered. For example, many results known for the lattice Polycon(n) (such as Hadwiger's characterization theorem) remain open questions for the ostensibly simpler lattice $\mathcal{P}^n$ of *polytopes* in $\mathbf{R}^n$, as well as for the lattice of polyconvex subsets of the sphere $\mathbf{S}^n$ (see also Chapter 11). For a survey of open questions in this area, see [71, 72].

Another lattice sharing many qualities with Polycon(n) is the lattice of *star-shaped* sets in $\mathbf{R}^n$. A set $A \subseteq \mathbf{R}^n$ is said to be *star-shaped*, if $0 \in A$, and if for each line ℓ passing through the origin in $\mathbf{R}^n$, the set $A \cap \ell$ is a closed interval. Advances have recently been made in the development of a comprehensive, though far from complete, theory of star-shaped sets, a theory *dual* to the theory of convex sets developed in the present text. In the dual theory, convex bodies are replaced by star-shaped sets, and projections onto subspaces are replaced by intersections with subspaces. In [64] Lutwak introduced dual mixed volumes, in analogy to the classical mixed volumes of Minkowski (see also [65]). Dual (star-shaped set) analogues also exist for the intrinsic volumes, the mean projection formula (replaced in the dual theory by a mean *intersection* formula; see [65]), kinematic formulas [103], and Hadwiger's characterization theorem [51, 52]. A comprehensive introduction to the theory of star-shaped sets and geometric tomography has been presented in Gardner's book [26].

Finally, one can ask whether an analogous lattice theory is possible for the collection of finite unions of compact convex subsets of the Grassmannians Gr(n, k), for which the notion of 'convex set' is suitably defined. Indeed, one may go even further, to the study of polyconvex subsets of compact Lie groups, about which almost nothing is known at present. A characterization theorem in the spirit of Hadwiger's characterization theorem for invariant valuations on compact Lie groups would have profound consequences throughout mathematics.

10

Kinematic formulas for polyconvex sets

In this chapter we use Hadwiger's characterization Theorem 9.1.1 to generalize the kinematic formulas of Chapter 3 to the lattice of polyconvex sets. These formulas will have application to additional questions regarding random motions of polyconvex sets. In Section 10.2 we use the principal kinematic formula of Section 10.1 to give a condition under which one compact convex set in $\mathbf{R}^2$ must be contained in a translate of another.

10.1 The principal kinematic formula

Before pursuing additional applications of Theorem 9.1.1, we digress briefly to discuss the Haar measure on the group of Euclidean motions (including reflections across hyperplanes). It is easy to derive an explicit formula for such a measure by the arguments we have already employed for Grassmannians. The Haar measure on the orthogonal group $O(n)$ is closely related to the invariant measure on the set of frames (and, therefore, the set of flags of subspaces) in $\mathbf{R}^n$, since orthogonal transformations are parametrized by Cartesian coordinate systems. In other words, after fixing a Cartesian coordinate system $x_1, x_2, \ldots, x_n$ in $\mathbf{R}^n$, an arbitrary orthogonal transformation is uniquely determined by the choice of another Cartesian coordinate system $u_1, u_2, \ldots, u_n$ with the same origin. (See also Section 6.7.)

Therefore, an invariant measure on $O(n)$ is obtained from the invariant measure on the set of frames in $\mathrm{Mod}(n)$ after multiplying it by 2^n. However, for the purposes of this section we shall normalize the invariant measure on $O(n)$ so that the total measure of $O(n)$ equals one; i.e. we use the Haar *probability* measure on $O(n)$. Recall that E_n denotes the set of all Euclidean motions of $\mathbf{R}^n$. Since every Euclidean motion is a (possibly improper) rotation followed by a translation, an invariant measure is induced on E_n by taking the product of the Haar probability measure on $O(n)$ with the n-dimensional Lebesgue measure. We denote this Haar measure on E_n by $\mathrm{d}g$, where $g \in E_n$.

It turns out that Sylvester's Theorem 7.2.3 can be generalized to the case in which, in place of a linear variety, we substitute any compact convex set. Let A and K be compact convex sets. For $g \in E_n$, denote by gK the set $\{g(b) : b \in K\}$, and consider the integral

$$\mu_0(A, K) = \int_{E_n} \mu_0(A \cap gK)\,\mathrm{d}g. \tag{10.1}$$

This integral has an evident geometric interpretation as the measure of the set of all $g \in E_n$ such that $A \cap gK \neq \emptyset$. Alternatively one may think of (10.1) as the 'measure' of all convex sets gK in $\mathbf{R}^n$ congruent to K that meet A. If A and C are both compact convex sets of dimension n and if $C \supseteq A$, then the quotient

$$\frac{\int \mu_0(A \cap gK)\,\mathrm{d}g}{\int \mu_0(C \cap gK)\,\mathrm{d}g} \tag{10.2}$$

gives the probability that a 'rigid' convex set K, dropped at random on $\mathbf{R}^n$ so as to meet C, shall also meet A. Note that this probability is independent of the choice of normalizing constant for the Haar measure $\mathrm{d}g$.

The computation of the probability (10.2) can be carried out using Hadwiger's Theorem 9.1.1. Indeed, the set function $\mu_0(A, K)$ is a continuous valuation in each of its variables K and A when the other is held fixed. Moreover,

$$\mu_0(g'A, K) = \int \mu_0(g'A \cap gK)\,\mathrm{d}g = \int \mu_0(A \cap g'^{-1}gK)\,\mathrm{d}g$$
$$= \int \mu_0(A \cap gK)\,\mathrm{d}g$$

by the invariance of the Euler characteristic and of Haar measure. One verifies similarly that $\mu_0(A, g'K) = \mu_0(A, K)$ for all $g' \in E_n$. Finally, note that

$$\int \mu_0(A \cap gK)\,\mathrm{d}g = \int \mu_0(g^{-1}A \cap K)\,\mathrm{d}g = \int \mu_0(gA \cap K)\,\mathrm{d}g^{-1}$$
$$= \int \mu_0(gA \cap K)\,\mathrm{d}g,$$

since the Haar measure on E_n is invariant under the inversion map $g \mapsto$

g^{-1}. Thus, $\mu_0(A, K) = \mu_0(K, A)$. From two applications of Hadwiger's Theorem 9.1.1 we obtain the expansion

$$\mu_0(A, K) = \sum_{i=0}^{n} \sum_{j=0}^{n} c_{ij} \mu_i(A) \mu_j(K),$$

where the coefficients $c_{ij} = c_{ji}$ depend only on n, and where μ_i and μ_j denote the intrinsic volumes.

It turns out that most of the constants c_{ij} are equal to zero. In fact, they can even be computed explicitly.

Theorem 10.1.1 (The principal kinematic formula) *For all* $A, K \in$ Polycon(n),

$$
\begin{aligned}
\int_{E_n} \mu_0(A \cap gK)\, \mathrm{d}g &= \sum_{i=0}^{n} \binom{n}{i}^{-1} \frac{\omega_i \omega_{n-i}}{\omega_n} \mu_i(A) \mu_{n-i}(K) \\
&= \sum_{i=0}^{n} \begin{bmatrix} n \\ i \end{bmatrix}^{-1} \mu_i(A) \mu_{n-i}(K).
\end{aligned}
$$

Proof From the argument above we know that

$$\mu_0(A, K) = \int_{E_n} \mu_0(A \cap gK)\, \mathrm{d}g = \sum_{i,j=0}^{n} c_{ij} \mu_i(A) \mu_j(K),$$

where the c_{ij} are constants depending only on i, j, and n, and where $c_{ij} = c_{ji}$.

Let B_n denote the unit ball in $\mathbf{R}^n$, centered at the origin. For $a, b \geq 0$, denote by aB_n and bB_n the balls of radii a and b in $\mathbf{R}^n$, centered at the origin. Then

$$
\begin{aligned}
\mu_0(aB_n, bB_n) &= \int_{E_n} \mu_0(aB_n \cap gbB_n)\, \mathrm{d}g \\
&= \int_{O(n)} \int_{\mathbf{R}^n} \mu_0(aB_n \cap (\phi bB_n + v))\, \mathrm{d}v\, \mathrm{d}\phi \\
&= \int_{O(n)} \int_{\mathbf{R}^n} \mu_0(aB_n \cap (bB_n + v))\, \mathrm{d}v\, \mathrm{d}\phi,
\end{aligned}
$$

since $\phi B_n = B_n$ for any orthogonal transformation ϕ. Since the total measure of $O(n)$ is equal to one, we continue:

$$\mu_0(aB_n, bB_n) = \int_{\mathbf{R}^n} \mu_0(aB_n \cap (bB_n + v))\, \mathrm{d}v$$

$$= \int_{\mathbf{R}^n} I_{(a+b)B_n} \, dv$$

$$= (a+b)^n \omega_n$$

$$= \omega_n \sum_{i=0}^{n} \binom{n}{i} a^i b^{n-i}.$$

In other words, $\mu_0(aB_n, bB_n)$ is a homogeneous polynomial in the non-negative variables a and b. Meanwhile,

$$\mu_0(aB_n, bB_n) = \sum_{i,j=0}^{n} c_{ij}\mu_i(aB_n)\mu_j(bB_n)$$

$$= \sum_{i,j=0}^{n} c_{ij} a^i b^j \mu_i(B_n)\mu_j(B_n).$$

Therefore, $c_{ij} = 0$ if $i + j \neq n$, and

$$c_{i,n-i} = \frac{\omega_n}{\mu_i(B_n)\mu_{n-i}(B_n)}\binom{n}{i} = \binom{n}{i}^{-1}\frac{\omega_i \omega_{n-i}}{\omega_n} = \begin{bmatrix} n \\ i \end{bmatrix}^{-1},$$

by Theorem 9.2.4. $\qquad\qquad\qquad\qquad\qquad\qquad\qquad\qquad\qquad\qquad\square$

In view of Theorem 10.1.1, the expression (10.2) becomes

$$\frac{\int \mu_0(A \cap gK) \, dg}{\int \mu_0(C \cap gK) \, dg} = \frac{\sum_{i=0}^{n} \begin{bmatrix} n \\ i \end{bmatrix}^{-1} \mu_i(A)\mu_{n-i}(K)}{\sum_{i=0}^{n} \begin{bmatrix} n \\ i \end{bmatrix}^{-1} \mu_i(C)\mu_{n-i}(K)},$$

giving once again the probability that a 'rigid' convex set K, dropped at random on $\mathbf{R}^n$ so as to meet C, shall also meet A.

The principal kinematic formula can also be expressed in more probabilistic terms, using random variables. For a compact convex set K in $\mathbf{R}^n$ let $X_i(K)$ again denote the i-volume of a projection of K onto a randomly chosen i-dimensional subspace $V \in \mathrm{Gr}(n,i)$. Combining Corollary 9.4.2 with the principal kinematic formula (Theorem 10.1.1) we obtain

$$\int_{E_n} \mu_0(K \cap gL) \, dg = \sum_{i=0}^{n} \begin{bmatrix} n \\ i \end{bmatrix} E(X_i(K))E(X_{n-i}(L)),$$

a formula mysteriously reminiscent of the classical binomial theorem.

10.2 Hadwiger's containment theorem

As an application of the principal kinematic formula 10.1.1, we derive Hadwiger's condition for the containment of one planar convex region by another. Theorem 10.1.1 takes the following form for polyconvex subsets of $\mathbf{R}^2$:

$$\int_{E_2} \mu_0(K \cap gL)\, \mathrm{d}g = \sum_{i=0}^{2} \begin{bmatrix} 2 \\ i \end{bmatrix}^{-1} \mu_i(K)\mu_{2-i}(L)$$

$$= \begin{bmatrix} 2 \\ 0 \end{bmatrix}^{-1} \mu_0(K)\mu_2(L) + \begin{bmatrix} 2 \\ 1 \end{bmatrix}^{-1} \mu_1(K)\mu_1(L) + \begin{bmatrix} 2 \\ 2 \end{bmatrix}^{-1} \mu_2(K)\mu_0(L),$$

for all $K, L \in \mathcal{K}^2$. The formula (6.4) for flag coefficients then yields

$$\int_{E_2} \mu_0(K \cap gL)\, \mathrm{d}g = \mu_0(K)\mu_2(L) + \frac{2}{\pi}\mu_1(K)\mu_1(L) + \mu_2(K)\mu_0(L).$$

$$(10.3)$$

Now suppose that K and L are convex *polygons* in $\mathbf{R}^2$ with non-empty interiors. Denote by ∂K and ∂L the boundaries of K and L respectively, and let g be a Euclidean motion of the plane. If $K \cap gL = \emptyset$ then $\partial K \cap g \partial L = \emptyset$ as well. However, if $K \cap gL \neq \emptyset$ there are two possibilities.

The first possibility is that $\partial K \cap g\partial L \neq \emptyset$ as well, in which case $\partial K \cap g\partial L$ consists of an even number of distinct points in $\mathbf{R}^2$. (There is a possibility that ∂K and $g\partial L$ will intersect in some other way, but the set of such motions g is easily seen to be of measure zero in E_2.)

The second possibility is that $\partial K \cap g\partial L$ is still empty, which implies that either $K \subseteq \mathrm{int}\ gL$ or $gL \subseteq \mathrm{int}\ K$. In other words, the motion g moves one of the regions K or L into the interior of the other.

Suppose that neither K nor L can be moved into the interior of the other by any rigid motion of the plane. This eliminates the second possibility. In terms of the Euler characteristic, this non-containment assumption implies that

(i) If $\mu_0(K \cap gL) = 0$ then $\mu_0(\partial K \cap g\,\partial L) = 0$ as well, and

(ii) If $\mu_0(K \cap gL) = 1$ then $\mu_0(\partial K \cap g\,\partial L) = 2k$ for some positive integer k,

for all $g \in E_2$ (except a set of measure zero). In other words, if neither K nor L can be moved into the interior of the other by any rigid motion of the plane, then we have

$$\mu_0(\partial K \cap g\,\partial L) \geq 2\mu_0(K \cap gL).$$

It follows that

$$\int_{E_2} \mu_0(\partial K \cap g \partial L) \, dg \geq \int_{E_2} 2\mu_0(K \cap gL) \, dg.$$

The principal kinematic formula (10.3) for $\mathbf{R}^2$ then implies that

$$\mu_0(\partial K)\mu_2(\partial L) \ + \ \frac{2}{\pi}\mu_1(\partial K)\mu_1(\partial L) + \mu_2(\partial K)\mu_0(\partial L)$$

$$\geq \ 2\mu_0(K)\mu_2(L) + \frac{4}{\pi}\mu_1(K)\mu_1(L) + 2\mu_2(K)\mu_0(L).$$

Because K and L are convex, we have $\mu_0(K) = \mu_0(L) = 1$. Since K and L are polygons, ∂K and ∂L consist of finite unions of line segments. Therefore, $\mu_2(\partial K) = \mu_2(\partial L) = 0$, and we have

$$\frac{2}{\pi}\mu_1(\partial K)\mu_1(\partial L) \geq 2\mu_2(L) + \frac{4}{\pi}\mu_1(K)\mu_1(L) + 2\mu_2(K). \qquad (10.4)$$

If we denote by $A(K)$ and $P(K)$ the area and perimeter of a compact convex subset K of $\mathbf{R}^2$ (with a non-empty interior), the inequality (10.4) becomes

$$\frac{2}{\pi}P(K)P(L) \geq 2A(L) + \frac{4}{\pi}\frac{P(K)}{2}\frac{P(L)}{2} + 2A(K),$$

so that

$$2\pi(A(K) + A(L)) - P(K)P(L) \leq 0. \qquad (10.5)$$

Given $K, L \in \mathcal{K}^2$ with non-empty interiors, define $\Delta(K, L)$ by

$$\Delta(K, L) = 2\pi(A(K) + A(L)) - P(K)P(L). \qquad (10.6)$$

We have shown that, if K and L are convex *polygons* with non-empty interiors, and if neither K nor L can be moved into the interior of the other by any rigid motion of the plane, then $\Delta(K, L) \leq 0$. Since the set of convex polygons is dense in the set of all compact convex subsets of $\mathbf{R}^2$, a simple continuity argument implies that $\Delta(K, L) \leq 0$ if $K, L \in \mathcal{K}^2$ (not necessarily polygons) such that neither contains a translate of the other in its interior. Thus, we have the following theorem.

Theorem 10.2.1 (Hadwiger's containment theorem) *Let $K, L \in \mathcal{K}^2$ with non-empty interiors. If $\Delta(K, L) > 0$ then there exists a Euclidean motion $g \in E_2$ such that either $K \subseteq \text{int } gL$ or $L \subseteq \text{int } gK$.*
$\square$

Which way will the containment go? If $\Delta(K, L) > 0$ then Theorem 10.2.1 implies that $A(K) \neq A(L)$. Therefore, the convex body of larger area will contain a rigid motion of the other.

10.3 Higher kinematic formulas

In addition to the principal kinematic formula 10.1.1 one can derive analogous kinematic formulas for the remaining intrinsic volumes $\mu_1, \ldots, \mu_n$. Define a function ζ_k on pairs of polyconvex sets by the formula

$$\zeta_k(A, K) = \int_{E_n} \mu_k(A \cap gK) \, dg$$

For *convex* A and K, one may interpret the expression

$$\frac{\zeta_k(A, K)}{\zeta_0(A, K)} = \frac{\int_{E_n} \mu_k(A \cap gK) \, dg}{\int_{E_n} \mu_0(A \cap gK) \, dg}$$

as the mean value, or the *expected value*, of the kth intrinsic volume of $A \cap gK$, taken over all gK in $\mathbf{R}^n$ congruent to K that meet A. If A and C are compact convex sets of dimension n and if $C \supseteq A$, then the quotient

$$\frac{\int \mu_k(A \cap gK) dg}{\int \mu_0(C \cap gK) dg} \tag{10.7}$$

gives the expected value of $\mu_k(A \cap gK)$ given that K meets C.

For A, K convex, Theorem 9.3.1 asserts that

$$\mu_k(A \cap gK) = \lambda_{n-k}^n(\mathrm{Graff}(A \cap gK; n-k))$$

Therefore, we have

$$
\begin{aligned}
\zeta_k(A, K) &= \int_{E_n} \mu_k(A \cap gK) \, dg \\
&= \int_{E_n} \lambda_{n-k}^n(\mathrm{Graff}(A \cap gK; n-k)) \, dg \\
&= \int_{E_n} \int_{\mathrm{Graff}(n,n-k)} \mu_0((A \cap gK) \cap V) \, d\lambda_{n-k}^n(V) \, dg \\
&= \int_{\mathrm{Graff}(n,n-k)} \int_{E_n} \mu_0((A \cap V) \cap gK) \, dg \, d\lambda_{n-k}^n(V) \\
&= \int_{\mathrm{Graff}(n,n-k)} \sum_{i=0}^{n} \begin{bmatrix} n \\ i \end{bmatrix}^{-1} \mu_i(A \cap V) \mu_{n-i}(K) \, d\lambda_{n-k}^n(V) \\
&= \sum_{i=0}^{n-k} \begin{bmatrix} n \\ i \end{bmatrix}^{-1} \mu_{n-i}(K) \int_{\mathrm{Graff}(n,n-k)} \mu_i(A \cap V) \, d\lambda_{n-k}^n(V) \\
&= \sum_{i=0}^{n-k} \begin{bmatrix} i+k \\ k \end{bmatrix} \begin{bmatrix} n \\ i \end{bmatrix}^{-1} \mu_{n-i}(K) \mu_{i+k}(A),
\end{aligned}
$$

where the last equality follows from Theorem 9.3.2. Hence, we have the following generalization of Theorem 10.1.1 for the intrinsic volumes.

Theorem 10.3.1 (The general kinematic formula) *For* $0 \leq k \leq n$,

$$\int_{E_n} \mu_k(A \cap gK) \, dg = \sum_{i=0}^{n-k} \begin{bmatrix} i+k \\ k \end{bmatrix} \begin{bmatrix} n \\ i \end{bmatrix}^{-1} \mu_{k+i}(A) \mu_{n-i}(K).$$

for all $A, K \in \mathrm{Polycon}(n)$. $\qquad \square$

Theorem 10.3.1 and Hadwiger's characterization Theorem 9.1.1 together provide a kinematic formula for the integral

$$\int_{E_n} \mu(A \cap gK) \, dg,$$

for any continuous invariant valuation μ on $\mathrm{Polycon}(n)$.

10.4 Notes

The principal kinematic formula 10.1.1 has variously been attributed to Blaschke, Chern, Hadwiger, and Santaló (see [82, p. 262] and [85, p. 253]). In [46], Howard gave a generalization of the kinematic formula to Riemannian homogeneous spaces. By a different approach Fu proved a generalized principal kinematic formula for invariant differential forms on homogeneous spaces [25]. Zhang [103] recently developed dual kinematic formulas for the lattice of star-shaped sets in $\mathbf{R}^n$. For numerous variations and applications of kinematic formulas, see also [82, 85, 87, 98, 106].

Hadwiger's containment Theorem 10.2.1 first appeared in [35, 36]; see also [82, pp. 121–123]. A spherical version of Hadwiger's containment theorem is treated in Section 11.4; see also [82, p. 324]. The question of how to generalize Theorem 10.2.1 to higher dimensions remains open in many cases; some new ideas have recently been put forward by Zhou [77, 104, 105, 106, 107] for convex bodies in $\mathbf{R}^3$ and in $\mathbf{R}^{2n}$, and by Zhang [102] for $\mathbf{R}^n$ in general. For a generalization of Theorem 10.2.1 to the projective and hyperbolic planes, see [31].

Once again we emphasize the analogical nature of these results. In particular, the kinematic formula 3.2.5 of Chapter 3 is a discrete analogue of the general kinematic formula 10.3.1 for $\mathrm{Polycon}(n)$. This analogy suggests the possibility of additional kinematic formulas for the lattice $\mathrm{Mod}(n)$, with applications to appropriate probabilistic questions in this context.

11

Polyconvex sets in the sphere

The successful characterization of many classes of valuations on polytopes and compact convex sets in Euclidean space motivates similar
questions about valuations on polytopes and convex sets in non-Euclidean spaces. Unfortunately little is yet known about valuations in a
non-Euclidean context. In this chapter we consider valuations on the
lattice of polyconvex sets in the unit sphere $\mathbf{S}^n$, with emphasis on the
elementary example of continuous $O(3)$-invariant valuations on the unit
sphere $\mathbf{S}^2$ in $\mathbf{R}^3$.

11.1 Convexity in the sphere

Recall that $\mathbf{S}^n$ denotes the set of unit vectors in $\mathbf{R}^{n+1}$. The intersection
$E \cap \mathbf{S}^n$ of a two-dimensional subspace E of $\mathbf{R}^{n+1}$ with the sphere is
called a *great circle*. More generally, the set $E \cap \mathbf{S}^n$ is called a *great
k-subsphere* of $\mathbf{S}^n$ if E is a subspace of $\mathbf{R}^{n+1}$ of dimension $k + 1$. Two
points $x, y, \in \mathbf{S}^{n-1}$ are said to be *antipodal* if $y = -x$.

The sphere $\mathbf{S}^n$ inherits a Riemannian structure from the ambient
space $\mathbf{R}^{n+1}$, in which the shortest path (or geodesic) between two non-
antipodal points $x \neq y \in \mathbf{S}^n$ is given by the shorter arc of the unique
great circle in $\mathbf{S}^n$ passing through x and y. A great $(n-1)$-subsphere σ
separates $\mathbf{S}^n$ into two *hemispheres*, each the antipode of the other. The
intersection of $n + 1$ distinct hemispheres having linearly independent
normals is called a *spherical n-simplex*. The intersection of at most n
hemispheres is called a *lune*.

For $u \in \mathbf{S}^n$ denote by $u^\perp$ the n-dimensional subspace of $\mathbf{R}^{n+1}$ orthogonal to u. If Δ is a simplex inside the great $(n-1)$-subsphere $u^\perp \cap \mathbf{S}^n$,
the *lune through* Δ, denoted $L(\Delta)$, consists of the union of all half circles
with endpoints at u and $-u$ and bisected by a point of Δ.

A set $P \subseteq \mathbf{S}^n$ is a *convex spherical polytope* if P can be expressed as a finite intersection of hemispheres. Denote by $\mathcal{P}(\mathbf{S}^n)$ the set of all convex spherical polytopes in $\mathbf{S}^n$. A *spherical polytope* is a finite union of convex spherical polytopes.

More generally, a set $K \subseteq \mathbf{S}^n$ will be called *convex* if K is contained in some hemisphere of $\mathbf{S}^n$ and if any two points of K can be connected by a geodesic (i.e. an arc of a great circle) inside K. Alternatively, a set K lying inside a hemisphere of $\mathbf{S}^n$ is convex if the cone $o * K$ in $\mathbf{R}^{n+1}$ defined by

$$o * K = \{\lambda u : u \in K \text{ and } 0 \le \lambda \le 1\}$$

is convex in $\mathbf{R}^{n+1}$.

Denote by $\mathcal{K}(\mathbf{S}^n)$ the set of all *compact* convex sets in $\mathbf{S}^n$. The set $\mathcal{K}(\mathbf{S}^n)$ is endowed with the topology induced by the *Hausdorff metric* on compact sets in $\mathbf{R}^{n+1}$ (see Section 4.1). In analogy to the Euclidean case, we call a set $K \subseteq \mathbf{S}^n$ *polyconvex* if K can be expressed as a finite union of compact convex sets in $\mathbf{S}^n$. Evidently the collection of polyconvex sets in $\mathbf{S}^n$ forms a distributive lattice under union and intersection of sets.

A function $\varphi : \mathcal{P}(\mathbf{S}^n) \longrightarrow \mathbf{R}$ is called a *valuation* on $\mathcal{P}(\mathbf{S}^n)$ if $\varphi(\emptyset) = 0$, where $\emptyset$ is the empty set, and if

$$\varphi(P \cup Q) = \varphi(P) + \varphi(Q) - \varphi(P \cap Q), \qquad (11.1)$$

for all $P, Q \in \mathcal{P}(\mathbf{S}^n)$ such that $P \cup Q \in \mathcal{P}(\mathbf{S}^n)$ as well. A valuation φ on $\mathcal{P}(\mathbf{S}^n)$ is *simple* if it vanishes on spherical polytopes of dimension less than n. A valuation φ on $\mathcal{P}(\mathbf{S}^n)$ will be called *invariant* if $\varphi(gP) = \varphi(P)$ for all orthogonal transformations (rotations and reflections) g of $\mathbf{S}^n$.

Similarly, a function $\varphi : \mathcal{K}(\mathbf{S}^n) \longrightarrow \mathbf{R}$ is said to be a valuation if $\varphi(\emptyset) = 0$ and if (11.1) is satisfied for all compact convex sets $K, L \subseteq \mathbf{S}^n$ such that $K \cup L \in \mathcal{K}(\mathbf{S}^n)$ as well. The following is an adaptation of Groemer's extension Theorem 5.1.1 for valuations on the sphere $\mathbf{S}^n$.

Theorem 11.1.1 (Groemer's extension theorem for $\mathbf{S}^n$)

  (i) *A valuation φ defined on convex polytopes in $\mathbf{S}^n$ admits a unique extension to a valuation on the lattice of all polytopes in $\mathbf{S}^n$.*

  (ii) *A continuous valuation φ on compact convex sets in $\mathbf{S}^n$ admits a unique extension to a valuation on the lattice of polyconvex sets in $\mathbf{S}^n$.*

In each case the extension of φ to finite unions is given by iteration of the inclusion–exclusion identity (11.1). $\qquad\square$

The proof of part (i) of Theorem 11.1.1 follows the same lines as the proof of Theorem 4.1.3. Similarly, the proof of Theorem 5.1.1 (for the Euclidean case) goes through for the sphere $\mathbf{S}^n$ without essential change, since the original proof is based not on the geometry of $\mathbf{R}^n$, but rather on the algebra of indicator functions, and on the fact that a polytope is the intersection of half spaces in $\mathbf{R}^n$, a property which carries over analogously to spherical polytopes and hemispheres. (See Section 5.1.) In the arguments that follow, the unique extension of a valuation φ given by Theorem 11.1.1 will allow us to consider the value of φ on all finite unions of convex spherical polytopes (or compact spherical convex sets in $\mathbf{S}^n$), whether or not such unions are actually convex.

11.2 A characterization for spherical area

We now turn our attention to the two-dimensional sphere $\mathbf{S}^2$. Important examples of continuous invariant valuations on $\mathcal{P}(\mathbf{S}^2)$ (and $\mathcal{K}(\mathbf{S}^2)$) include *spherical area*, denoted φ_2, *spherical length* φ_1, and the *Euler characteristic* φ_0.

The spherical length $\varphi_1(K)$ of a spherical convex region K with a non-empty interior in $\mathbf{S}^2$ is given by one half of the perimeter of K; that is, one half of the length of the curve in $\mathbf{S}^2$ that forms the boundary of K. It is easy to verify that φ_1 is an extension of geodesic length in $\mathbf{S}^2$ to a continuous invariant valuation on $\mathcal{K}(\mathbf{S}^2)$. To see this, note that

$$\varphi_1(K) = 2\mu_2(o * K) - \varphi_2(K),$$

for all $K \in \mathcal{K}(\mathbf{S}^2)$.

The Euler characteristic $\varphi_0(K)$ of a spherical convex region K is defined to be 1 if $K \neq \emptyset$, while $\varphi_0(\emptyset) = 0$. Theorem 11.1.1 insures that φ_0 has a unique continuous and invariant extension to all finite unions of spherical convex sets in K. As in the Euclidean case, this unique extension of φ_0 coincides with the Euler characteristic of algebraic topology.

It can be shown that in fact every continuous invariant valuation on $\mathcal{P}(\mathbf{S}^2)$ (or $\mathcal{K}(\mathbf{S}^2)$) is a linear combination of the valuations $\varphi_0, \varphi_1, \varphi_2$. To this end, we prove a characterization theorem for spherical area φ_2.

Theorem 11.2.1 (The spherical area theorem) *Suppose that φ is a continuous invariant simple valuation on $\mathcal{P}(\mathbf{S}^2)$. Then there exists $c \in \mathbf{R}$ such that $\varphi(P) = c\varphi_2(P)$, for all $P \in \mathcal{P}(\mathbf{S}^2)$.*

Remark: Theorem 11.2.1 actually holds under the weaker assumption that φ is invariant under the group $SO(3)$ of *rotations* of $\mathbf{S}^2$. This follows from an argument similar to that in the proof of Proposition 8.3.1.

Before proving Theorem 11.2.1, we consider two preliminary cases.

Proposition 11.2.2 *Suppose that φ is a continuous invariant simple valuation on closed arcs I of the circle $\mathbf{S}^1$. Then there exists $c \in \mathbf{R}$ such that $\varphi(I) = c\varphi_1(I)$, for all closed arcs $I \subseteq \mathbf{S}^1$.*

Proof Let $c = \varphi(\mathbf{S}^1)/(2\pi)$, and define $\nu(I) = \varphi(I) - c\varphi_1(I)$, for all closed arcs $I \subseteq \mathbf{S}^1$. Note that ν is also continuous, invariant, and simple. In addition, $\nu(\mathbf{S}^1) = 0$. It now suffices to show that $\nu(I) = 0$ for all I.

Suppose that I_n is a closed arc of length $2\pi/n$, where n is a positive integer. Since the circle $\mathbf{S}^1$ can be tiled with n rotations of I_n, the invariance and simplicity of ν imply that $n\nu(I_n) = \nu(\mathbf{S}^1) = 0$, so that $\nu(I_n) = 0$. Since any closed arc I of length $2\pi m/n$ can be tiled with rotations of I_n, for any positive integers $m < n$, it follows that ν vanishes on arcs of length rational in proportion to the circle. It then follows from the continuity of ν that $\nu(I) = 0$ for all closed arcs I. $\quad\square$

Proposition 11.2.3 *Suppose that φ is a continuous invariant simple valuation on $\mathcal{P}(\mathbf{S}^2)$ such that $\varphi(\mathbf{S}^2) = 0$. Then $\varphi(\Delta) = 0$, for all spherical simplices $\Delta \subset \mathbf{S}^2$.*

Proof Let σ denote a great circle in $\mathbf{S}^2$. For any closed arc I contained in a half-circle in σ, denote by $L(I)$ the lune through I, and define $\nu(I) = \varphi(L(I))$. Evidently the valuation ν satisfies the conditions of Proposition 11.2.2. Therefore, there exists $c \in \mathbf{R}$ such that $\nu(I) = c\varphi_1(I)$ for all $I \subseteq \sigma$. However, $\nu(\sigma) = \varphi(\mathbf{S}^2) = 0$, so that $c = 0$. It follows that φ vanishes on all lunes through closed arcs of σ. Since φ is invariant, it follows in turn that φ vanishes on all lunes in $\mathbf{S}^2$.

Now suppose that Δ is a spherical 2-simplex in $\mathbf{S}^2$; that is, a spherical triangle; given by the intersection of hemispheres $\Delta = H_1 \cap H_2 \cap H_3$. See Figure 11.1. For $A \subseteq \mathbf{S}^2$ denote by A^c the closure of the complement $\mathbf{S}^2 - A$. Note that

$$(H_1 \cup H_2 \cup H_3)^c = H_1^c \cap H_2^c \cap H_3^c = -\Delta,$$

where $-\Delta$ denotes the orthogonal reflection of Δ through the origin in $\mathbf{R}^3$. Since φ is simple and invariant, we obtain

$$\varphi(H_1 \cup H_2 \cup H_3) = \varphi(\mathbf{S}^2) - \varphi((H_1 \cup H_2 \cup H_3)^c) = 0 - \varphi(-\Delta) = -\varphi(\Delta). \tag{11.2}$$

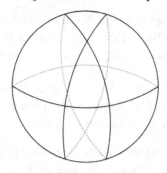

Fig. 11.1. An intersection of hemispheres in $\mathbf{S}^2$.

Meanwhile, iteration of the inclusion–exclusion identity (11.1) gives

$$\varphi(H_1 \cup H_2 \cup H_3) = \sum_{i=1}^{3} \varphi(H_i) - \sum_{i<j} \varphi(H_i \cap H_j) + \varphi(H_1 \cap H_2 \cap H_3). \ (11.3)$$

Note that the intersections $H_i \cap H_j$ are lunes in $\mathbf{S}^2$ for $i \neq j$. Since φ vanishes on hemispheres and lunes, the identity (11.3) implies that

$$\varphi(H_1 \cup H_2 \cup H_3) = \varphi(H_1 \cap H_2 \cap H_3) = \varphi(\Delta). \qquad (11.4)$$

It then follows from (11.2) and (11.4) that $\varphi(\Delta) = -\varphi(\Delta)$; that is, $\varphi(\Delta) = 0$. $\qquad\qquad\qquad\qquad\qquad\qquad\qquad\qquad\qquad\qquad\qquad\qquad\qquad\quad$ □

Remark: A spherical triangle Δ in $\mathbf{S}^2$ has *angle measures* $\alpha, \beta, \gamma \in [0, \pi]$ if the inner Euclidean (dihedral) angles between planar sides of the cone $o * \Delta$ in $\mathbf{R}^3$ are given by α, β, γ. It is well known that the spherical area of Δ is given by the *spherical excess*:

$$\varphi_2(\Delta) = \alpha + \beta + \gamma - \pi. \qquad (11.5)$$

One can prove (11.5) by the same inclusion–exclusion technique that was used in the proof of Proposition 11.2.3, using the fact that the lune through a great circular arc of length θ has spherical area 2θ. We are now ready to prove Theorem 11.2.1.

Proof of Theorem 11.2.1 Suppose that φ is a continuous invariant simple valuation on $\mathcal{P}(\mathbf{S}^2)$. Let $c = \varphi(\mathbf{S}^2)/(4\pi)$ and define $\nu(P) = \varphi(P) - c\varphi_2(P)$ for all $P \in \mathcal{P}(\mathbf{S}^2)$. Since the valuation ν satisfies the conditions of Proposition 11.2.3, it follows that $\nu(\Delta) = 0$ for all spherical simplices $\Delta \subseteq \mathbf{S}^2$.

For $P \in \mathcal{P}(\mathbf{S}^2)$ express P as a union of spherical simplices

$$P = \Delta_1 \cup \cdots \cup \Delta_m,$$

where $\dim(\Delta_i \cap \Delta_j) < 2$ for all $i \neq j$. Since ν is simple, it follows that

$$\nu(P) = \sum_{i=1}^{m} \nu(\Delta_i) = 0,$$

for all $P \in \mathcal{P}(\mathbf{S}^2)$. □

Since any continuous invariant simple valuation φ on $\mathcal{K}(\mathbf{S}^2)$ restricts to such a valuation on the *dense* subspace $\mathcal{P}(\mathbf{S}^2)$ of convex spherical polytopes, Theorem 11.2.1 and the continuity of φ immediately imply the following theorem.

Theorem 11.2.4 *Suppose that φ is a continuous invariant simple valuation on $\mathcal{K}(\mathbf{S}^2)$. Then there exists $c \in \mathbf{R}$ such that $\varphi(K) = c\varphi_2(K)$, for all $K \in \mathcal{K}(\mathbf{S}^2)$.* □

11.3 Invariant valuations on spherical polytopes

Theorem 11.2.1 leads to the following characterization theorem for continuous invariant valuations.

Theorem 11.3.1 *Suppose that φ is a continuous invariant valuation on $\mathcal{P}(\mathbf{S}^2)$. Then there exist $c_0, c_1, c_2 \in \mathbf{R}$ such that, for all $P \in \mathcal{P}(\mathbf{S}^2)$,*

$$\varphi(P) = c_0\varphi_0(P) + c_1\varphi_1(P) + c_2\varphi_2(P).$$

Theorem 11.3.1 can be thought of as a spherical analogue of Hadwiger's characterization Theorem 9.1.1. Since the set $\mathcal{P}(\mathbf{S}^2)$ is dense in $\mathcal{K}(\mathbf{S}^2)$, Theorem 11.3.1 also holds if $\mathcal{P}(\mathbf{S}^2)$ is replaced with the larger collection $\mathcal{K}(\mathbf{S}^2)$.

Proof Suppose that φ is a continuous invariant valuation on $\mathcal{P}(\mathbf{S}^2)$. Let $x \in \mathbf{S}^2$, and let $c_0 = \varphi(\{x\})$. Since φ is invariant, the value of c_0 is independent of the choice of x. Therefore, the valuation $\varphi - c_0\varphi_0$ vanishes on singleton sets in $\mathbf{S}^2$.

If we restrict $\varphi - c_0\varphi_0$ to a great circle $\sigma \subset \mathbf{S}^2$, then by Proposition 11.2.2 there exists $c_1 \in \mathbf{R}$ such that $\varphi - c_0\varphi_0 = c_1\varphi_1$ on $\mathcal{P}(\sigma)$. Once again it follows from the invariance of φ, φ_0, and φ_1 that the real number c_1 is a constant independent of our choice of the great circle σ. It follows that the valuation ν on $\mathcal{P}(\mathbf{S}^2)$ given by

$$\nu = \varphi - c_0\varphi_0 - c_1\varphi_1$$

vanishes on all $P \in \mathcal{P}(\mathbf{S}^2)$ of dimension less than 2; that is, ν is a continuous invariant simple valuation on $\mathcal{P}(\mathbf{S}^2)$. Theorem 11.2.1 then implies the existence of $c_2 \in \mathbf{R}$ such that $\nu(P) = c_2\varphi_2(P)$ for all $P \in \mathcal{P}(\mathbf{S}^2)$. $\qquad\qquad\qquad\qquad\qquad\qquad\qquad\qquad\qquad\qquad\qquad\qquad$ □

Theorem 11.3.1 implies that any continuous invariant valuation on $\mathcal{P}(\mathbf{S}^2)$ is determined completely by its values on a singleton $\{x\}$, a great circle σ, and the entire sphere $\mathbf{S}^2$. Indeed, the coefficients c_i in Theorem 11.3.1 can be computed using the following table of values:

	$\{x\}$	σ	$\mathbf{S}^2$
φ_0	1	0	2
φ_1	0	2π	0
φ_2	0	0	4π

$$(11.6)$$

Note well that $\varphi_1(\mathbf{S}^2) = 0$. This follows from the fact that $\varphi_1(H) = \pi$, for any closed hemisphere H, while $\varphi_1(\sigma) = 2\pi$ if σ is a great circle. Denote by $-H$ the antipode of H. We then obtain

$$\varphi_1(\mathbf{S}^2) = \varphi_1(H \cup -H) = \varphi_1(H) + \varphi_1(-H) - \varphi_1(H \cap -H) = 0,$$

since $H \cap -H$ is a great circle.

One of the advantages of Hadwiger's characterization Theorem 9.1.1 for invariant valuations on Euclidean convex sets is the ease with which that theorem allows one to prove a variety of classical results in integral geometry. We now demonstrate similar advantages of the spherical area Theorem 11.2.1 and the resultant characterization Theorem 11.3.1.

We begin with a spherical analogue of the mean projection formula (9.7) of Section 9.4. For $K \in \mathcal{K}(\mathbf{S}^2)$ and $u \in \mathbf{S}^2$, define the *projection* K_u of K onto the great circle $u^\perp \cap \mathbf{S}^2$ by

$$K_u = \{x \in u^\perp \cap \mathbf{S}^2 : [ux] \cap K \neq \emptyset\},$$

where $[ux]$ denotes the (unique) half great circle through x with endpoints at u and $-u$.

The *mean spherical width* sw(K) of a compact convex set K in $\mathbf{S}^2$ is defined by

$$\mathrm{sw}(K) = \int_{\mathbf{S}^2} \varphi_1(K_u) \, du. \qquad\qquad (11.7)$$

The functional sw(K) is continuous on $\mathcal{K}(\mathbf{S}^2)$. To see this, note that, for a fixed $u \in \mathbf{S}^2$ and $K \in \mathcal{K}(\mathbf{S}^2)$, the interval K_u varies continuously in a neighborhood of (K, u) provided that either $u \in \mathrm{int}(K)$ or $u \notin K$. Thus the functional $\varphi_1(K_u)$ is continuous at (K, u) unless $u \in \partial K$. Since ∂K

has measure zero in $\mathbf{S}^2$, the functional $\varphi_1(K_u)$ is integrable with respect to spherical Lebesgue measure, and the integral (11.7) varies continuously with respect to K. Although we do *not* apply the formula (11.7) to non-convex sets, it is possible to extend the functional sw to a continuous valuation on all polyconvex sets in $\mathbf{S}^2$ as follows.

Suppose that $K \subset \mathbf{S}^2$ is compact and convex. If $u \in \mathbf{S}^2$ and $x \in u^\perp \cap \mathbf{S}^2$, then $\varphi_0(K \cap [ux]) = 1$ if and only if $x \in K_u$, otherwise $\varphi_0(K \cap [ux]) = 0$. (An exception may occur if K contains a great half-circle, in which case this exception will occur with measure zero.) Therefore, the equation (11.7) can be rewritten

$$\mathrm{sw}(K) = \int_{\mathbf{S}^2} \varphi_1(K_u)\, du = \int_{\mathbf{S}^2} \int_{u^\perp \cap \mathbf{S}^2} \varphi_0(K \cap [ux])\, dx\, du. \quad (11.8)$$

Since φ_0 is an invariant valuation, it follows from (11.8) that the mean spherical width $\mathrm{sw}(K)$ is a continuous invariant valuation of K. Moreover, the integral (11.8) is defined for all polyconvex sets in $\mathbf{S}^2$, giving the desired extension of sw.

Proposition 11.3.2 *For $K \in \mathcal{K}(\mathbf{S}^2)$,*

$$\mathrm{sw}(K) = 4\pi \varphi_1(K) + 2\pi \varphi_2(K).$$

Proof Since sw is a continuous invariant valuation on $\mathcal{K}(\mathbf{S}^2)$, Theorem 11.3.1 implies that

$$\mathrm{sw}(K) = c_0 \varphi_0(K) + c_1 \varphi_1(K) + c_2 \varphi_2(K),$$

for some constants $c_i \in \mathbf{R}$. Since the formula (11.8) is valid (as a valuation) for all finite unions of compact convex sets, we can evaluate (11.8) to compute sw on great circles and on the full sphere $\mathbf{S}^2$. To compute the coefficients c_i, note that sw vanishes on singleton sets, so that $c_0 = 0$. If σ is a great circle, then $\sigma \cap [ux]$ consists of a single point *almost always*; that is, provided that $u \notin \sigma$. Thus, $\varphi_0(\sigma \cap [ux]) = 1$ almost always, and we obtain

$$c_1 \varphi_1(\sigma) = 2\pi c_1 = \int_{\mathbf{S}^2} \int_{u^\perp \cap \mathbf{S}^2} dx\, du = 8\pi^2,$$

or $c_1 = 4\pi$. Finally, if $K = \mathbf{S}^2$ then $\mathbf{S}^2 \cap [ux] = [ux]$, a half-circle, so that $\varphi_0(\mathbf{S}^2 \cap [ux]) = 1$. Since $\varphi_1(\mathbf{S}^2) = 0$, we obtain

$$c_2 \varphi_2(\mathbf{S}^2) = 4\pi c_2 = \int_{\mathbf{S}^2} \int_{u^\perp \cap \mathbf{S}^2} dx\, du = 8\pi^2,$$

or $c_2 = 2\pi$. $\qquad \square$

In a similar manner Theorem 11.3.1 can be used to obtain an easy proof of the spherical Crofton formula, which expresses the spherical perimeter (or length) of a spherical polyconvex set K as the measure of the great circles meeting K; that is,

$$\int_{\mathbf{S}^2} \varphi_0(K \cap u^\perp) \, du = 4\varphi_1(K), \qquad (11.9)$$

for all polyconvex sets K in $\mathbf{S}^2$.

11.4 Spherical kinematic formulas

In analogy to the methods of Sections 10.1 and 10.3, the characterization Theorem 11.3.1 yields an easy proof of a *spherical* kinematic formula for polyconvex sets in $\mathbf{S}^2$, which generalizes the spherical Crofton formula (11.9).

Theorem 11.4.1 (The principal kinematic formula for $\mathbf{S}^2$) *For all polyconvex sets $K, L \subseteq \mathbf{S}^2$,*

$$\int_{O(3)} \varphi_0(K \cap gL) \, dg \;=\; \frac{1}{4\pi}\varphi_0(K)\varphi_2(L) + \frac{1}{2\pi^2}\varphi_1(K)\varphi_1(L)$$
$$+ \frac{1}{4\pi}\varphi_2(K)\varphi_0(L) - \frac{1}{8\pi^2}\varphi_2(K)\varphi_2(L).$$

Here the integral is taken with respect to the invariant Haar probability measure on the orthogonal group $O(3)$.

Proof To begin, define

$$\varphi_0(K, L) = \int_{O(3)} \varphi_0(K \cap gL) \, dg.$$

For fixed K, the set function $\varphi_0(K, L)$ is a valuation in the variable L; in fact, it is an invariant valuation, since

$$\varphi_0(K, g_0 L) \;=\; \int_{O(3)} \varphi_0(K \cap gg_0 L) \, dg$$
$$=\; \int_{O(3)} \varphi_0(K \cap gL) \, dg,$$

for each $g_0 \in O(3)$. By Theorem 11.3.1, the functional $\varphi_0(K, L)$ can be expressed as a linear combination of the valuations φ_i, with coefficients

$c_i(K)$ depending on K:

$$\varphi_0(K, L) = \sum_{i=0}^{2} c_i(K)\varphi_i(L).$$

Meanwhile, for fixed L, the set function $\varphi_0(K, L)$ is a valuation in the variable K. From this it follows that each of the coefficients $c_i(K)$ is a valuation in the variable K. One way to check this fact is to insert special values for L and use the table (11.6). Since φ_0 is invariant, we have

$$
\begin{aligned}
\varphi_0(K, L) &= \int_{O(3)} \varphi_0(K \cap gL) \, \mathrm{d}g \\
&= \int_{O(3)} \varphi_0(g^{-1}K \cap L) \, \mathrm{d}g \\
&= \int_{O(3)} \varphi_0(gK \cap L) \, \mathrm{d}g = \varphi_0(L, K).
\end{aligned}
$$

Therefore, the coefficients $c_i(K)$ are invariant valuations in the parameter K, so that

$$\varphi_0(K, L) = \sum_{i,j=0}^{2} c_{ij}\varphi_i(K)\varphi_j(L), \qquad (11.10)$$

again by Theorem 11.3.1. Since $\varphi_0(K, L) = \varphi_0(L, K)$, it is evident that $c_{ij} = c_{ji}$. One can verify this, as well as compute the values of the coefficients c_{ij}, by evaluating $\varphi_0(K, L)$ for the cases in which each of K and L is either a point (singleton) $\{x\}$, a great circle σ, or the whole sphere $\mathbf{S}^2$, using the values of φ_i given in the table (11.6).

For example, if $K = L = \{x\}$, then (11.10) takes the form:

$$0 = c_{00}.$$

We then let $K = \sigma$ and $L = \{x\}$, so that (11.10) becomes

$$
\begin{aligned}
0 &= c_{01}\varphi_0(\sigma)\varphi_1(\{x\}) + c_{10}\varphi_1(\sigma)\varphi_0(\{x\}) + c_{11}\varphi_1(\sigma)\varphi_1(\{x\}) \\
&= 0 + 2\pi c_{10} + 0
\end{aligned}
$$

so that $c_{10} = c_{01} = 0$.

In a similar manner one evaluates both sides of (11.10) for each case of $K, L \in \{\{x\}, \sigma, \mathbf{S}^2\}$ (using the table (11.6)) to compute the values of c_{ij}, thereby completing the proof of Theorem 11.4.1. $\qquad \Box$

Using similar methods one can easily show that, for polyconvex sets $K, L \subseteq \mathbf{S}^2$,

$$\int_{O(3)} \varphi_1(K \cap gL) \, \mathrm{d}g = \frac{1}{4\pi}\varphi_1(K)\varphi_2(L) + \frac{1}{4\pi}\varphi_2(K)\varphi_1(L),$$

and

$$\int_{O(3)} \varphi_2(K \cap gL) \, \mathrm{d}g = \frac{1}{4\pi}\varphi_2(K)\varphi_2(L),$$

thereby classifying all kinematic formulas for continuous invariant valuations on $\mathbf{S}^2$ (by Theorem 11.3.1).

Theorem 11.4.1 and its higher dimensional generalizations have numerous applications to questions in spherical geometric probability. For example, Theorem 11.4.1 (as stated for $\mathbf{S}^2$) leads in turn to a spherical analogue of Hadwiger's containment Theorem 10.2.1.

Theorem 11.4.2 *Let $K, L \in \mathcal{K}(\mathbf{S}^2)$ with non-empty interiors, and suppose that*

$$2\pi(\varphi_2(K) + \varphi_2(L)) - \varphi_2(K)\varphi_2(L) - 4\varphi_1(K)\varphi_1(L) > 0. \qquad (11.11)$$

Then there exists $g \in O(3)$ such that either $K \subseteq \mathrm{int} \, gL$ or $L \subseteq \mathrm{int} \, gK$.
□

To determine the direction of containment in Theorem 11.4.2, one compares the values of $\varphi_2(K)$ and $\varphi_2(L)$. Evidently the set of larger spherical area will contain a rigid motion of the other.

If we define $A(K) = \varphi_2(K)$ and $P(K) = 2\varphi_1(K)$ (spherical area and perimeter respectively), then the inequality condition (11.11) becomes:

$$2\pi(A(K) + A(L)) - A(K)A(L) - P(K)P(L) > 0.$$

Compare this with (10.5) in Section 10.2.

In analogy to the proof of Theorem 10.2.1, the proof of Theorem 11.4.2 makes use of the spherical principal kinematic formula (Theorem 11.4.1), along with the fact that two polygonal convex curves in the sphere will almost always intersect at an even number of points. For details, see Section 10.2.

11.5 Remarks on higher dimensional spheres

A natural question at this point (or even earlier) is that of how spherical volume on polytopes or convex sets in higher dimensional spheres is to be characterized. In order to characterize spherical area (on $\mathbf{S}^2$) we proved the following sequence of assertions for a given continuous invariant simple valuation φ:

$$\varphi \text{ is simple} \mapsto \varphi = \text{area on } \textit{lunes} \text{ (by induction on dimension)}$$
$$\mapsto \varphi = \text{spherical area,}$$

where the second implication follows from an inclusion–exclusion argument (see (11.3)). Unfortunately, this inclusion–exclusion argument fails in $\mathbf{S}^3$ and for $\mathbf{S}^{2n+1}$ in general. Without a characterization theorem for spherical volume in $\mathbf{S}^3$, the induction step for the first implication is not possible, so that the proof of Theorem 11.2.1 also fails to generalize to $\mathbf{S}^4$ and so on. However, we do have the following partial result.

Denote by S the (invariant) spherical volume on $\mathbf{S}^{2n}$. Recall that a *lune* in $\mathbf{S}^{2n}$ is a subset of $\mathbf{S}^{2n}$ consisting of the intersection of at most $2n$ hemispheres.

Theorem 11.5.1 *Suppose that φ is a continuous invariant simple valuation on $\mathcal{P}(\mathbf{S}^{2n})$. If $\varphi(L) = S(L)$ for all lunes $L \subseteq \mathbf{S}^{2n}$, then $\varphi(P) = S(P)$ for all $P \in \mathcal{P}(\mathbf{S}^{2n})$.*

Proof Without loss of generality we may assume that $\varphi(\mathbf{S}^{2n}) = 0$, so that $\varphi(L) = 0$ for all lunes $L \subseteq \mathbf{S}^{2n}$. (Just substitute $\varphi - S$ for φ.) It remains to show that $\varphi(P) = 0$ for all $P \in \mathcal{P}(\mathbf{S}^{2n})$.

Suppose that Δ is a spherical simplex in $\mathbf{S}^{2n}$ given by the intersection of hemispheres $\Delta = H_1 \cap \cdots \cap H_{2n+1}$. Again denote by A^c the closure of the complement $\mathbf{S}^{2n} - A$, for any $A \subseteq \mathbf{S}^{2n}$. Note that

$$\left(\bigcup_{i=1}^{2n+1} H_i \right)^c = \bigcap_{i=1}^{2n+1} H_i^c = -\Delta.$$

Since φ is simple and invariant, we obtain

$$\varphi \left(\bigcup_{i=1}^{2n+1} H_i \right) = \varphi(\mathbf{S}^{2n}) - \varphi \left(\left(\bigcup_{i=1}^{2n+1} H_i \right)^c \right) = 0 - \varphi(-\Delta) = -\varphi(\Delta).$$
$$(11.12)$$

Meanwhile, the inclusion–exclusion principle gives

$$\varphi \left(\bigcup_{i=1}^{2n+1} H_i \right) = \sum_{i=1}^{2n+1} \varphi(H_i) - \sum_{i_1 < i_2} \varphi(H_{i_1} \cap H_{i_2}) + \cdots$$
$$+ (-1)^{2n} \varphi(H_1 \cap \cdots \cap H_{2n+1}). \qquad (11.13)$$

Since φ vanishes on hemispheres and lunes, all of the terms on the right-hand side of (11.13) vanish except for the last term: $(-1)^{2n}\varphi(H_1 \cap \cdots \cap H_{2n+1}) = \varphi(\Delta)$. It then follows from (11.12) that

$$\varphi(\Delta) = \varphi\left(\bigcup_{i=1}^{2n+1} H_i\right) = -\varphi(\Delta),$$

so that $\varphi(\Delta) = 0$. Since every polytope $P \in \mathcal{P}(\mathbf{S}^{2n})$ can be expressed as a union of spherical simplices intersecting in dimension less that $2n$, it follows that $\varphi(P) = 0$ for all $P \in \mathcal{P}(\mathbf{S}^{2n})$. □

This proof of Theorem 11.5.1 fails for $\mathbf{S}^{2n+1}$ because the sign of the last term on the right-hand side of (11.13) is negative in this case, so that, on combining it with (11.12), we would obtain the tautological observation that

$$\varphi(\Delta) = -\varphi\left(\bigcup_{i=1}^{2n+1} H_i\right) = \varphi(\Delta).$$

For this reason a proof of such a spherical volume characterization for $\mathbf{S}^{2n+1}$, if possible, will require some new ingredient besides the decomposition of simplices by inclusion–exclusion. Indeed the question of whether spherical volume is the only continuous invariant simple valuation on $\mathcal{P}(\mathbf{S}^n)$ (or $\mathcal{K}(\mathbf{S}^n)$) remains open for dimension $n \geq 3$.

11.6 Notes

Theorems 11.2.1 and 11.5.1 are both originally due to McMullen. In [70] McMullen developed the *polytope algebra*, which places many dissection and inclusion–exclusion techniques in an algebraic framework (see also [72]). The proofs of Theorems 11.2.1 and 11.5.1 given in Section 11.2 give a more simplified approach to these specific questions.

Groemer made observations regarding Theorem 11.1.1 at the end of his paper on extensions of additive set functionals [32]. Theorem 11.4.2 and other non-Euclidean analogues of Hadwiger's containment theorem can be found in [82, p. 324]. For additional references see Section 10.4.

Theorem 11.4.1 generalizes to a principal kinematic formula for spheres of arbitrary dimension; see [82, p. 321]. Similarly, the spherical Crofton formula (11.9) generalizes to higher dimensions [82, p. 316]. However, it is not known whether the characterization Theorem 11.3.1 generalizes to spheres of dimension $n \geq 3$.

Theorems 11.2.1 and 11.2.4 vastly simplify valuation characterization theorems for valuations on *star-shaped* sets in $\mathbf{R}^3$. In [51, 52] characterizations are given for continuous homogeneous valuations and continuous rotation invariant valuations on a class of star-shaped sets called L^n-*stars*; that is, star-shaped sets in $\mathbf{R}^n$ with n-integrable radial functions (see also [26]). For the general case of star-shaped sets in $\mathbf{R}^n$ these characterizations require that the valuations in question be defined on the entire class of L^n-stars. However, by using Theorem 11.2.4 one can prove similar characterizations for valuations defined merely on finite unions of convex cones in $\mathbf{R}^3$ (of finite height) having a common apex at the origin. This is a much smaller and more manageable class of objects, which is nonetheless dense in the original and larger class of L^3-stars. Clearly a higher dimensional version of Theorem 11.2.1 or Theorem 11.2.4 would have similar and important consequences in the theory of valuations on star-shaped sets in $\mathbf{R}^n$.

Bibliography

[1] R. V. Ambartzumian. *Combinatorial Integral Geometry*. (New York: John Wiley and Sons, 1982).

[2] R. V. Ambartzumian. *Factorization Calculus and Geometric Probability*. (New York: Cambridge University Press, 1990).

[3] I. Anderson. *Combinatorics of Finite Sets*. (New York: Oxford University Press, 1987).

[4] E. Artin. *The Gamma Function*. (New York: Holt, Rinehart, and Winston, Inc., 1964).

[5] E. Barbier. Note sur le problème de l'aiguille et le jeu du joint couvert, *J. Mathematiques Pures et Appliquées (2)*, **5** (1860), 273–286.

[6] C. Berge. Sur une propriété combinatoire des ensembles convexes, *C. R. Acad. Sci. Paris*, **248** (1959), 2698–2699.

[7] C. Berge. *Hypergraphs*. (New York: Elsevier Science Publishers, 1989).

[8] V. Boltianskii. *Hilbert's Third Problem*. (New York: John Wiley and Sons, 1978).

[9] A. Cauchy. Note sur divers théorèms relatifs à la rectification des courbes et à la quadrature des surfaces, *C. R. Acad. Sci., Paris*, **13** (1841), 1060–1065.
Also: *Oeuvres Complètes (1)*, **6** (Paris: 1888), pp. 369–375.

[10] A. Cauchy. Mémoire sur la rectification des courbes et la quadrature des surfaces courbes, *Mém. Acad. Sci. Paris*, **22** (1850), 3.
Also: *Oeuvres Complètes (1)*, **2** (Paris: 1908), pp. 167–177.

[11] B. Chen, The Gram–Sommerville and Gauss–Bonnet theorems and geometric measures for noncompact polyhedra, *Adv. Math.*, **91** (1992), 269–291.

[12] B. Chen, On the Euler characteristic of finite unions of convex sets, *Disc. Comput. Geom.*, **10** (1993), 79–93.

[13] B. Chen and G.-C. Rota. Totally invariant set functions of polynomial type, *Comm. Pure Appl. Math.*, **47** (1994), 187–197.

[14] M. W. Crofton. On the theory of local probability, applied to straight lines drawn at random in a plane; the methods used being also extended

to the proof of certain new theorems in the integral calculus, *Philos. Trans. Roy. Soc. London*, **158** (1868), 181–199.

[15]   M. W. Crofton. Sur quelques théorèms de calcul intégral, *C. R. Acad. Sci., Paris*, **68** (1869), 1469–1470.

[16]   M. W. Crofton. Geometrical theorems relating to mean values, *Proc. London Math. Soc.*, **8** (1877), 304–309.

[17]   M. W. Crofton. Probability. *Encyclopaedia Britannica*, 9th Edition, **19** (1885), 768–788.

[18]   L. Danzer, B. Grünbaum and V. Klee. Helly's theorem and its relatives. *Convexity, Proceedings of Symposia in Pure Mathematics*, Volume VII. (Providence, RI: American Mathematical Society, 1963), 101–177.

[19]   M. Dehn. Über den Rauminhalt, *Math. Ann.*, **55** (1901), 465–478.

[20]   G. Dolzmann and D. Hug. Equality of two representations of extended affine surface area, *Arch. Math.*, **65** (1995), 352–356.

[21]   J. Eckhoff. Helly, Radon, and Caratheódory type theorems. *Handbook of Convex Geometry*, Peter M. Gruber and Jörg M. Wills, Eds. (Amsterdam: North-Holland, 1993), 389–448.

[22]   P. Erdős. On a lemma of Littlewood and Offord, *Bull. Amer. Math. Soc.*, **51** (1945), 898–902.

[23]   W. Feller. *An Introduction to Probability Theory and Its Applications*, Volume II, 2nd Edition. (New York: Wiley Publishers, 1957).

[24]   S. Fisk. Whitney numbers of projective space over **R**, **C**, **H**, and the p-adics, *J. Combinatorial Theory, Series A*, **70** (1995), 165–169.

[25]   J. H. G. Fu. Kinematic formulas in integral geometry. *Indiana Univ. Math. J.*. **39** (1990), 1115–1154.

[26]   R. J. Gardner. *Geometric Tomography* (New York: Cambridge University Press, 1995).

[27]   *Classic Papers in Combinatorics,* Ira Gessel and G.-C. Rota, Eds. (Boston: Birkhäuser Verlag, 1987).

[28]   A. Ghouila-Houri. Sur l'étude combinatoire des familles de convexes, *C. R. Acad. Sci. Paris*, **252** (1961), 494–496.

[29]   C. Greene and D. Kleitman. Proof techniques in the theory of finite sets. *Studies in Combinatorics*, G.-C. Rota, Ed. (Mathematical Association of America, 1978), 22–79.

[30]   N. T. Gridgeman. Geometric probability and the number π, *Scripta Math.*, **25** (1960), 183–195.

[31]   E. Grinberg, D. Ren and J. Zhou. The symmetric isoperimetric deficit and the containment problem in a plane of constant curvature. Preprint.

[32]   H. Groemer. On the extension of additive functionals on classes of convex sets, *Pacific J. Math.*, **75** (1978), 397–410.

[33]   H. Groemer. Fourier series and spherical harmonics in convexity. *Handbook of Convex Geometry*, Peter M. Gruber and Jörg M. Wills, Eds. (Amsterdam: North-Holland, 1993), 1259–1296.

[34]   H. Groemer. *Geometric Applications of Fourier Series and Spherical Harmonics.* (New York: Cambridge University Press, 1996).

170   *Bibliography*

[35] H. Hadwiger. Überdeckung ebener Bereiche durch Kreise und Quadrate, *Comment. Math. Helv.*, **13** (1941), 195–200.

[36] H. Hadwiger. Gegenseitige Bedeckbarkeit zweier Eibereiche und Isoperimetrie. *Vierteljschr. Naturforsch. Gesellsch. Zürich*, **86** (1941), 152–156.

[37] H. Hadwiger. Eulers Charakteristik und kombinatorische Geometrie, *J. reine angew. Math.*, **194** (1955), 101–110.

[38] H. Hadwiger. *Altes und Neues über Konvexe Körper*. (Basel: Birkhäuser Verlag, 1955).

[39] H. Hadwiger. *Vorlesungen über Inhalt, Oberfläche und Isoperimetrie.* (Berlin: Springer Verlag, 1957).

[40] H. Hadwiger and P. Glur. Zerlegungsgleichheit ebener Polygone, *Elem. Math.*, **6** (1951), 97–106.

[41] G. H. Hardy, J. E. Littlewood, and G. Pólya. *Inequalities*, 2nd Edition. (New York: Cambridge University Press, 1988).

[42] L. H. Harper and G.-C. Rota. Matching theory, an introduction, *Adv. Probability*, **1** (1971), 171–215.

[43] E. Helly. Über Mengen konvexer Körper mit gemeinschaftlichen Punkten, *Jber. Deutsch. Math. Verein.*, **32** (1923), 175–176.

[44] D. Hilbert. *Mathematical Problems*. Lecture delivered before the International Congress of Mathematicians in Paris, 1900. Translated by M. W. Newson, *Bull. Amer. Math. Soc.*, **8** (1902), 437–479.

[45] M. Hochberg and W. M. Hirsch. Sperner families, s-systems, and a theorem of Meshalkin, *Annals New York Acad. Sci.*, **175** (1970), 224–237.

[46] R. Howard. The kinematic formula in Riemannian homogeneous spaces, *Memoirs Amer. Math. Soc.*, **509** (1993).

[47] S. Karlin. *Total Positivity*. (Stanford, CA: Stanford University Press, 1968).

[48] G. Katona. A theorem of finite sets, *Theory of Graphs*, (Ákádémia Kiadó, Budapest: 1968), 187–207.

[49] M. Kendall and P. Moran. *Geometrical Probability*. (New York: Hafner Publishing Co., 1963).

[50] D. Klain. A short proof of Hadwiger's characterization theorem, *Mathematika*, **42** (1995), 329–339.

[51] D. Klain. Star valuations and dual mixed volumes, *Adv. Math.*, **121** (1996), 80–101.

[52] D. Klain. Invariant valuations on star-shaped sets, *Adv. Math.*, **125** (1997), 95–113.

[53] D. Klain. Kinematic formulas for finite vector spaces, to appear in *Discrete Math.*

[54] D. Klain. Even valuations on convex bodies. Preprint.

[55] D. Klain and G.-C. Rota. A continuous analogue of Sperner's theorem, *Comm. Pure Appl. Math.*, **50** (1997), 205–223.

[56] V. Klee. On certain intersection properties of convex sets, *Canadian J. Math.*, **3** (1951), 272–275.

[57] J. B. Kruskal. The number of simplices in a complex. *Mathematical Optimization Techniques.* (University of California Press, 1963), 251–278.

[58] T. Kubota. Über konvex-geschlossene Mannigfaltigkeiten im n-dimensionalen Raume, *Sci. Rep. Tôhoku Univ.*, **14** (1925), 85–99.

[59] K. Leichtweiss. Zur Affinoberfläche konvexer Körper, *Manuscripta Math.*, **56** (1986), 429–464.

[60] K. Leichtweiss. Über einige Eigenschaften der Affinoberfläche beliebiger konvexer Körper, *Resultate Math.*, **13** (1988), 255–282.

[61] K. Leichtweiss. Bemerkungen zur Definition einer erweiterten Affinoberfläche von E. Lutwak, *Manuscripta Math.*, **65** (1989), 181–197.

[62] F. W. Levi. Eine Ergänzung zum Hellyschen Satz, *Arch. Math.*, **4** (1953), 222–224.

[63] D. Lubell. A short proof of Sperner's theorem, *J. Combinatorial Theory*, 1 (1966), 299.

[64] E. Lutwak. Dual mixed volumes, *Pacific J. Math.*, **58** (1975), 531–538.

[65] E. Lutwak. Intersection bodies and dual mixed volumes, *Adv. Math.*, **71** (1988), 232–261.

[66] E. Lutwak. Extended affine surface area, *Adv. Math.*, **85** (1991), 39–68.

[67] E. Lutwak. Selected affine isoperimetric inequalities, *Handbook of Convex Geometry*, Peter M. Gruber and Jörg M. Wills, Eds. (Amsterdam: North-Holland, 1993), 151–176.

[68] E. Lutwak. Containment and circumscribing simplices, to appear in *Disc. Comput. Geom.*

[69] M. Marcus and H. Minc. *A Survey of Matrix Theory and Matrix Inequalities.* (New York: Dover Publications, 1992).

[70] P. McMullen. Non-linear angle-sum relations for polyhedral cones and polytopes, *Math. Proc. Camb. Phil. Soc.*, **78** (1975), 247–261.

[71] P. McMullen. Valuations and dissections, *Handbook of Convex Geometry*, Peter M. Gruber and Jörg M. Wills, Eds. (Amsterdam: North-Holland, 1993), 933–988.

[72] P. McMullen and R. Schneider. Valuations on convex bodies, *Convexity and Its Applications*, Peter M. Gruber and Jörg M. Wills, Eds. (Boston: Birkhäuser Verlag, 1983), 170–247.

[73] L. D. Meshalkin. A generalization of Sperner's theorem on the number of subsets of a finite set, *Theor. Probability Appl.*, **8** (1963), 203–204.

[74] J. Milnor and J. Stasheff. *Characteristic Classes* (Princeton, NJ: Princeton University Press, 1974).

[75] J. Munkres. *Elements of Algebraic Topology.* (Menlo Park, CA: Benjamin/Cummings, 1984).

[76] L. Nachbin. *The Haar Integral.* (New York: D. Van Nostrand Company, 1965).

[77] D. Ren. *Topics in Integral Geometry.* (Singapore: World Scientific International Publisher, 1992).

[78] G.-C. Rota. On the foundations of combinatorial theory, I: Theory of Möbius functions, *Z. Wahrscheinlichkeit.* 2, **368** (1964), 340–368.

[79]   G.-C. Rota. On the combinatorics of the Euler characteristic. *Studies in Pure Mathematics (Papers Presented to Richard Rado)* (London: Academic Press, 1971), 221–233.

[80]   *Gian-Carlo Rota on Combinatorics: Introductory Papers and Commentaries*, Joseph P. S. Kung, Ed. (Boston: Birkhäuser Verlag, 1995).

[81]   C.-H. Sah. *Hilbert's Third Problem: Scissors Congruence.* (San Francisco: Fearon Pitman Publishers, 1979).

[82]   L. A. Santaló. *Integral Geometry and Geometric Probability.* (Reading, MA: Addison-Wesley, 1976).

[83]   S. Schanuel. What is the length of a potato? An introduction to geometric measure theory. *Categories in Continuum Physics, 1982, Springer Lecture Notes in Math.*, **1174** (1986), 118–126.

[84]   S. Schanuel. Negative sets have Euler characteristic and dimension. *Category Theory, 1990, Springer Lecture Notes in Math.*, **1488** (1991), 379–385.

[85]   R. Schneider. *Convex Bodies: The Brunn–Minkowski Theory.* (New York: Cambridge University Press, 1993).

[86]   R. Schneider. Simple valuations on convex bodies, *Mathematika*, **43** (1996), 32–39.

[87]   R. Schneider and J.A. Wieacker. Integral geometry, *Handbook of Convex Geometry*, Peter M. Gruber and Jörg M. Wills, Eds. (Amsterdam: North-Holland, 1993), 1349–1390.

[88]   C. Schütt. On the affine surface area, *Proc. Amer. Math. Soc.*, **118** (1993), 1213–1218.

[89]   C. Schütt and E. Werner. The convex floating body, *Math. Scand.*, **66** (1990), 275–290.

[90]   H. Solomon. *Geometric Probability.* (Philadelphia: Society for Industrial and Applied Mathematics, 1978).

[91]   E. Sperner. Ein Satz über Untermengen einer endlichen Menge, *Math. Z.*, **27** (1928), 544–548.

[92]   R. Stanley. *Enumerative Combinatorics.* (Monterey, CA: Wadsworth & Brooks/Cole Advanced Books and Software, 1986).

[93]   E. Steinitz. Bedingt konvergente Reihen und konvexe Systeme, *J. reine angew. Math.*, **143** (1913), 128–175.

[94]   G. Strang. *Linear Algebra and its Applications*, 3rd Edition. (New York: Harcourt Brace Jovanovich, 1988).

[95]   J. J. Sylvester. On a funicular solution of Buffon's 'problem of the needle' in its most general form, *Acta Math.*, **14** (1890), 185–205.

[96]   R. Vitale. The Wills functional and Gaussian processes, *Ann. Probab.*, **24** (1996), 2172–2178.

[97]   F. W. Warner. *Foundations of Differentiable Manifolds and Lie Groups.* (New York: Springer Verlag, 1983).

[98]   W. Weil. Kinematic integral formulas for convex bodies, *Contributions to Geometry, Proc. Geometry Symp., Siegen, 1978*, J. Tölke and Jörg M. Wills, Eds. (Boston: Birkhäuser Verlag, 1978), 60–76.

[99]   W. Weil. Stereology: A survey for geometers, *Convexity and its Applications*, Peter M. Gruber and Jörg M. Wills, Eds. (Boston: Birkhäuser

Verlag, 1983), 360–412.

[100]   E. Werner. Illumination bodies and affine surface area, *Studia Math.*, **110** (1994), 257–269

[101]   J. M. Wills. Zur Gitterpunktanzahl konvexer Mengen, *Elemente Math.*, **28** (1973), 57–63.

[102]   G. Zhang. Geometric inequalities and inclusion measures of convex bodies, *Mathematika*, **41** (1994), 95–116.

[103]   G. Zhang. Dual kinematic formulas. Preprint.

[104]   J. Zhou. A kinematic formula and analogues of Hadwiger's theorem in space, *Contemporary Math.*, **140** (1992), 159–167.

[105]   J. Zhou. The sufficient condition for a convex body to fit another in $\mathbf{R}^4$, *Proc. Amer. Math. Soc.*, **121** (1994), 907–913.

[106]   J. Zhou. Kinematic formulas for mean curvature powers of hypersurfaces and Hadwiger's theorem in $\mathbf{R}^{2n}$, *Trans. Amer. Math. Soc.*, **345** (1994), 243–262.

[107]   J. Zhou. When can one domain enclose another in $\mathbf{R}^3$?, *J. Austral. Math. Soc. (Series A)*, **59** (1995), 266–272.

Index of symbols

Index

Printed in the United States
By Bookmasters